HANS-JOACHIM ZILLMER

The Energy Mistake

Figure Credits

Picture Credits

HANS-JOACHIM ZILLMER

The Energy Mistake

Why Crude Oil and Natural Gas are Exhaustless:
Plasma and the Electrical Solar System

Suppressed Facts,
Forbidden Proofs, Phony Dogmas

With 32 photos and 112 figures

The original **German Edition**:
"Der Energie-Irrtum. Warum Erdöl und Erdgas unerschöpflich sind."
Published 2009 in Munich for Germany, Swiss and Austria.
This German book is updated and expanded for publishing this English edition.

Translated in **foreign languages**: Czech (2010).

More information about the topics in this book: **www.zillmer.com**
Cover Image: Hans-Joachim Zillmer

Order this book online at www.trafford.com
or email orders@trafford.com

Most Trafford titles are also available at major online book retailers.

Printed in the United States of America.

ISBN: 978-1-4669-7276-6 (sc)
ISBN: 978-1-4669-7278-0 (e)

Library of Congress Control Number: 2012923566

Trafford rev. 01/22/2013

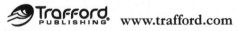 **www.trafford.com**

North America & international
toll-free: 1 888 232 4444 (USA & Canada)
phone: 250 383 6864 ♦ fax: 812 355 4082

Contents

1 Large Quantities of Carbon and Methane

False dogmas are blocking the development of human society. Humankind is at a dead end, and only a true world view can still save our future! Energy resources are increasingly becoming more expensive, allegedly because the natural reserves of prestigious fossil fuels will soon be depleted. The dramatic increase in prices of fossil fuels leads to shifts in the wealth of nations, and wars are consequently risked to ensure access to oil fields. All of these facts result from the prediction of a shortage of energy resources which are mixtures of hydrocarbons called "fossil fuels". However, spacecraft data from recent past years has provided evidence of the existence of hydrocarbons such as methane and ethane which exist in large amounts in our solar system. There evolvement is however not a consequence of biological processes. Nevertheless, hydrocarbons on Earth are still regarded as pure biological products.

Dinosaurs and Soft Coal

In my previous books, the fossilization of biological materials was controversially discussed. These processes of fossilisation can only happen within a relatively short period of time, but do not take place over millions of years, because biological material would be disintegrated during a fossilization process which is going on imperceptive slowly, with the precondition that an absolute exclusion of air was not present. Such fossilization processes have to evolve quickly, otherwise, fossilized eggs with fully preserved respectively non decomposed embryos inside or also fossilized droppings – so-called coprolites – would for example not be possible.

With these facts in mind let's have a look at fossils encased in black coal. In general, coal fossils represent "infusion fossils". This means that the structure of an organism has remained preserved, but its substance, however, was replaced to a large extent by liquids or gases which must have penetrated the biological structure. In principle, these fossils contain carbon with a percentage of about 90 – such as the coal itself. The thereby obtained fossil may be structurally almost perfect, although sometimes fiercely compressed, and not uncommon, under a microscope it is possible to clearly recognize fine details, even up to the cell structure. Nevertheless this biological structure is filled out with the same coal concentrate which is also surrounding the fossils.

The German botanist Henry Potonié (1905), based on a study of fossilized higher plants, concluded that black coal has a biological origin, "because the plant cells can be recognized immediately; and this without further special preparation" (ibid., p. 9). This was a turning point, as scientists previously believed; "the black coal is a mineral in the sense of something like quartz, feldspar, and mica, having originated in the same way" (ibid., p. 8). If, however, black coal had come into being in the same way as peat and lignite (ibid., p. 10), this would result in a coal–paradox.

Why does a single finely structured leaf of a tree remain preserved within a carbonaceous mass, whereas there is not a little bit left of the rest of the leaves? What is the reason that the branch or trunk belonging to the leaf had not remained? In structural terms, black coal therefore sharply differs from peat and brown coal (lignite), which certainly evolved from organic materials. How could such a coal-fossil evolve? Since the fine–segmented structure remained intact, today's homogeneous black coal must have once been a liquid or gas! Generally, a carbon or silicon containing mineral or fluid has to fill out the organic structure due to a kind of infusion process to start fossilization in this way. This chemical process must have taken place very quickly, since otherwise a leaf or even a fragile egg would have been rotten long time before the end of the process.

It is important to ascertain that petrified impressions of footprints can only be formed in soft mud layers, and not in the solid rock in which it can be found today. This soft mud which contain the footprints must have been hardened very quickly, like the prints of hands and foots celebrated in soft cement of the *Walk of Fame* in Beverly Hills, Los Angeles, because otherwise the prints would have been quickly destroyed by different erosion factors.

Fossilized footprints of dinosaurs were found at the ceiling of many coal mines in the Western United States – a still largely unknown phenomenon. In Utah, there are several coal mines in the vicinity of Helper and Price, In certain areas the soft peat surface had been heavily bioturbated by dinosaur activities with many footprints partly overlapping and obscuring previous tracks (e.g. Balsley / Parker, 1983, p. 279).

Also in other coal mines footprints of dinosaurs have been often documented, among others in the *Castlegate Mine* in the Rocky Mountains area (Peterson, 1924), in Wyoming, in the western part of Colorado, in Utah near Rock Springs and in New Mexico near Cuba (Gillette /Lockley, 1989). The very different sized footprints originate from carnivorous dinosaurs (theropods), officially regarded as bipedal dinosaurs – as well as four–legged herbivorous dinosaurs (sauropods), which roamed together in the same area. Isolated individual footprints are seldom documented, some of which are very large. The largest known to me is 1.36 meters long, but one has revealed to me a location that entails even larger ones. Interesting are, however, also 50 approximately 15-centimeter-long fossilized bird–like footprints that span across an area of approximately five square meters (Gillette / Lockley, 1989).

The previously described coal deposit that as a whole stretches from Wyoming over Utah and Colorado up to New Mexico belongs to the geologically very "young" black coal deposits, originated in the cretaceous age (Blackhawk Formation).

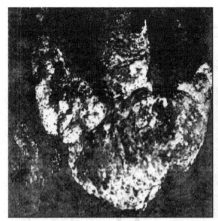

Figure 1: Coal-footprints. In the Castle Gate Mine there are toed Dinosaur footprints on the ceiling of the coal seams. Casts with footprints can be extracted from coal seam ceilings, in places where the dinosaurs sank (picture on the right hand side).

Normally black coal was created in the Carboniferous, a geological period ending 50 million years before the beginning of the Dinosaur era.

If black coal, according to conventional interpretation, should definitely have been originated out of organic material, it is surprising that some coal formations entail very few fossils or even none at all. Fossil free black coal can be found in Alaska for example. Overall wherever there are more fossils in the black coal, there are reported large variations in quantity. Also, the distribution of fossils in coal-seams itself is not homogeneous, because fossils are rare in the interior, but rather often found at or in other words *on* the ceiling of the coal seam. Therefore, the footprints of dinosaurs and birdlike animals are thus on and not within the coal seam.

These animals once walked on a thin layer of peat and sand of a fresh water swamp; underneath there was a still soft but hardening layer of coal. In this, dinosaurs sank down up to 30 centimeters deep. The feet were thereby vertically pulled out again as the shape of the fossilized footprints shows. There is evidence of long trails of footprints, whereby the dinosaurs mostly only sank after several steps. This is why castings exist in the seam.

Are softly coal seams a result of a uniform development respectively transformation which starts with swamps which become increasingly thicker and this slowly with time to emerge as denser black coal? On the contrary those coal seams were formed *relatively quickly* within a short timespan *as a whole*, because dinosaurs and bird-like animals could run over still soft coal layers? In the Kenilworth coal mine in Utah, fossilized footprints left by small animals were found on the ceiling of the coal layer. But it is amazing that exactly the same fossilized footprints were also found at the bottom of the coal seam

9

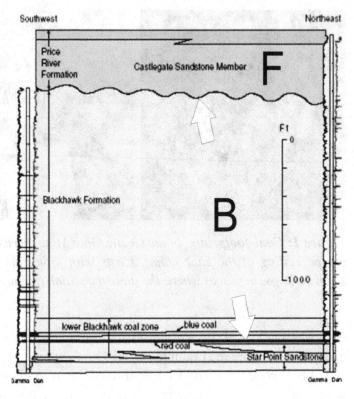

Fig. 2: Homogeneous coal. This section through a southern area of the Wasatch Plateau in Utah shows a homogeneous Blackhawk Coal Layer (B) with a thickness of over 400 metres. Located above are fluvial sandstone layers (F) deposited on the former Price River bed respectively on the coal seam. In this river bed, footprints were originated as dinosaurs swam in the river and walked on the river bed around. But also there are footprints in the sandstone layer immediately below the Blackhawk coal layer (arrows). In spite of that no footprints were discovered within the homogeneous coal layer (B).

or in other words *under* the coal seam (Gillette / Lockley, 1989). Therefore, this species must have thus existed at the beginning and also at the end of the coal development.

This brings us to the little–known fact that dinosaur prints were found not only on the ceiling, but also under the coal seams. These footprints are down there, even though not in the coal seam itself, but on the surface of the sediment layer which is located just beneath the coal layer and before hardening the *soft* carbonaceous substance – *not* swamp material – fills in these footprints. Thus the dinosaurs once walked on a soft sediment layer that consisted of sand in the over-area of a stream, however mixed with coal like components of the coal seam. These layers are located mostly in the upper part of the coal seam in the form of bulged humps caused by upwelling hydrocarbons. In contrast, there was no superimposed load and therefore not enough pressure to start coalification.

In the western part of North America on top of the sandstone layers – respectively under the coal seams – also fossilized remains of plants do exist and among others petrified fragments of palm trees in addition to footprints. It is believed that this sandstone was formed at the bottom of brackish water behind the coast lines (Gillette / Lockley, 1989). Consequently these coal layers were formed as a carbon deposit to its full height by a *uniformly* process *once* in a time period – which will to be discussed later on. Dinosaurs on the other hand walked around *on* the coal seam, where plentiful plant fossils can be found. On top of these coal seams there exist often thin layers of sediments, which must have been formed shortly one after another, because these layers were deformed *synchronistically* in an elastic–plastic state at once, because these today's hardened layers do not show any cracks. Due to their composition, it seems that these layers must have been quickly spread over each other, similar to what has been demonstrated during the outbreak of the volcano Mount St. Helens on 19 March 1982: At this time, numerous thin layers were formed on top of each other in a timespan of just a few hours and for example there was suddenly originated an eight meter thick, and over the entire height thinly banded geological formation (Zillmer, 2005, p. 114).

It is therefore technical possible, that foots of the dinosaurs had deformed *several* superimposed *soft* layers simultaneously.

We can draw the conclusion that the soft coal seam was formed during the lifetime of a particular species in its full height by a single process; since such animals existed prior and after the formation of the carbonaceous mass. The coal seam – like the sandstone layer underneath – was soft, so that dinosaurs partly sank into these layers. The carbon deposit with a thin cover of fluvial sediments forms the primeval sea or river floor in these regions – an almost *unknown fact that does not quite correspond to conventional theories.*

Now the question arises, how these previous soft coal seams have developed? Certainly not from plant and wood remains in the form of a peat bogs! In this case, one should also be able to find footprints and / or leaves in *different* heights of the coal layer.. According to the conventional theory, the initial organic material was transformed and compressed under high pressure, temperatures and the absence of air,. This process must had been lasted until a fixed composite of carbon was originated, in which water and fire–resistant forms of ash respectively charcoal were implemented. Indeed, sometimes large lumps of charcoal can be found within a compact black coal layer, what it needs air! But the coalification should take place under absence of air (Fig. 2a).

Also, the theory of coalification cannot explain that coal seams had built softly submarine grounds so that animals could walk over? But, because an assumed developing coalification process needs absence of air, it remains unclear why at the same time such a layer could built a submarine ground, on which animals were running around?

But also, there could force no great pressure which is supposedly required for a coalification process. The pressure was almost ridiculously low, because if dinosaurs and birds could left footprints in the submarine ground, this body of water was not very deep, and the sedimentary layers overlying the coal layer were still not existent at this time or very thin! Similarly, the required heat could hardly had evolved, because otherwise palm leaves and other plant parts would hardly have remain preserved: The only possibility is, this coal seams must have developed as low temperatures prevailed.

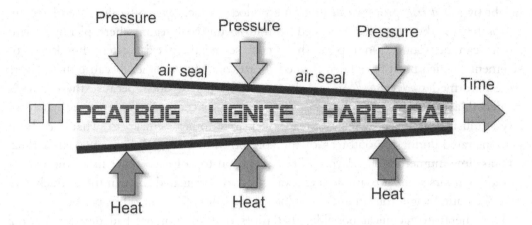

Pressure   Pressure   Pressure

air seal   air seal   Time

PEATBOG   LIGNITE   HARD COAL

Heat   Heat   Heat

Fig. 2a: Coalification process. *It is shown the general sequence of coalification from a peat bog to hard coal respectively anthracite. Since microbial activity shall ceases, the coalification process must be controlled primarily by changes in physical conditions that take place with depth, especially oxygen depletion due to airtightness and the important geochemical factors of higher pressure and temperature. In addition, some coal characteristics are assumed to originated at the time of peat formation — respectively charcoal–like material in coal is attributed to fires that occurred during dry periods while peat was still forming.*

Conclusion: *As evidenced in case of footprints of dinosaurs and birds, there was no thick sedimentary layer above the coal seam at this time. Accordingly the required pressure and temperature conditions and the required airtightness to start a process of coalification did not exist. Because hard coal could not develop in the way as conventional thought, this Cretaceous coal could not originate from organic material. Coalification did not take place! A new theory is absolutely necessary.*

Fig. 2b: Pre-coal seam. *This body, called The Nederfrederiksmose Body, discovered in 1898, is the first of such bodies to have been photographed at the site where it was found. As can be seen, a body sink to the bottom of a peat bog. Therefore, dinosaurs could not have walked around on a developing coal seam and therefore no tracks could occur on the surface of the seam. And sometime later, the pre–coal seam must been covered with thick sediment layers due to a coalification process.*

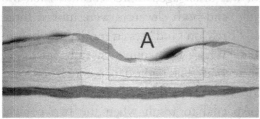

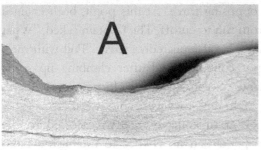

*Fig. 2c: **Deformed layers**. This fossilized dinosaur footprint shows that several layers have been deformed at the same time, as a dinosaur ran on muddy layers which are building not easily broken layers of rock these days.*

Even as seen from a biogeographically viewpoint, the plant material to build coal deposits cannot originate from tropical forests. Hoch much land area this trees should had covered in order to build the basis for the formation of all known coal seams? If we assume that a hectare land area was able to produce 700 solid cubic meters of wood. In such a wood mass, a certain amount of energy is included. Therefore we can figure out which area the forests would have had to overstretch that the resulting wood mass is matching the energy contained in all known coal deposits. As expected, the result is a huge area, twice as large as the current total land area. In this calculation we have assumed that *all* land areas were covered with forests, and that the wood deposits in their entirety would be buried airtight and that was ruling enough pressure and sufficient heat during the coalification process. Should all these "ideal" conditions not be present, the required land area accordingly increases.

But there will be still other controversial issues, because coal contains substances whose presence cannot be explained from a biological origin orientated perception. The fact that all coal and petroleum products contain radioactive uranium and up to four percent sulfur is not discussed. How these can occur worldwide in significant quantities in coal seams? Also why such deposits often contain a lot of methane, which would have to be degassed over the millions of years?

And Suddenly it Goes Down

Several mysterious plane crashes have occurred off the U.S. northeast coast. On 17th July 1996, the Trans World Airline Flight TWA 800 started to fly from JFK International Airport in New York to Paris. Only twelve minutes later the plane exploded, broke in half and crashed into the Sea, south of Long Island. All 230 passengers and crew members died. On 2nd September 1998, the Swissair Flight 111 was flying from the same airport heading northbound. 14 minutes after take-off both gadgets, the NY-TRACON (New York Terminal Radar Approach and Departure Control) and the NY-ARTCC (New York Air Route Traffic Control Center), lost radio contact with the plane, a little later a Canadian air traffic controller was able to establish radio contact with the plane. There was a report of in-flight smoke in the cockpit, before the Swissair McDonnell Douglas MD-11 impacted the Atlantic. All 215 passengers and 14 crew members were killed in the aircraft's collision with water.

Also at JFK International Airport the Egypt Air Flight 990 took-off for a regularly flight to Cairo in Egypt on 31st October 1999. Over international waters, about 100 kilometers south of Nantucket Island, Massachusetts, the Boeing 767-366ER aircraft autopilot disconnected. The cockpit voice recorder recorded the First Officer saying "I rely on God." A minute later, the autopilot was disengaged. Three seconds later, the throttles for both engines were reduced to idle, and both elevators were moved three degrees nose down. The First Officer repeated "I rely on God" seven more times before the Captain is suddenly heard to ask repeatedly, "What's happening, what's happening?" The plane dropped 14,600 feet (4,500 m) in 36 seconds The flight data recorder reflected that the elevators then moved into a split condition, with the left elevator up and the right elevator down; a condition which is expected to result when the two control columns are subjected to at least 50 lbs. of opposing force. At this point, both engines were shut down by moving the start levers from run to cutoff. The Captain asked, "What is this? What is this? Did you shut the engines?" and repeatedly stated, "Pull with me". But the elevator surfaces remained in a split condition and after climbing up some thousand feet the plane crashed into the Atlantic Ocean, killing all 217 people on board. The engines operated normally for the entire flight until they were shut down.

In the aforementioned cases a sudden event must have in each case led to the subsequent crash, at least in so far that the pilots had no chance to establish a radio contact. At the time of plane crashes in July 1996 and October 1999, gas flames and fireballs had been observed near the coast. Thomas Gold (1999) comes to the conclusion that the cause of these plane crashes may have been a result of those slight earthquakes that had released methane from the submarine ground. In this region, there are deposits of dense layers of methane hydrates (also called methane ice) of up to 500 meters height

14

on the ocean floor. Critics doubt whether methane gas can go through such thick layers of ice. But it is to take account of a large and *not* very small amount of methane. This gas is gushing out of the ocean floors and stored in freezing water if high water pressure is present to build a natural gas hydrate in which a large amount of methane is trapped within a crystal structure of water, forming a solid similar to crystalline ice.

The moment the methane ice layer attains a certain amount of denseness, it results in a build-up of methane gas under it, similar to that under a permafrost floor, for example in Siberia. The over-pressure in the rocks results in cracks in the upper oceanic crust, through which large amounts of gas can suddenly escape. In this case, the methane is not fully enclosed by water molecules, which means no methane ice is formed, so excessive gas leaks from the ocean floor. In any case, in a cubic meter of methane hydrate is more energy contained than in a cubic meter of natural gas, under the same pressure conditions. If the pressure deceases, the volume of methane gas trapped in the methane hydrates increases strongly, for example to 164-fold under the given pressure conditions in our atmosphere. This can lead to a "blow-out" when methane gas is explosively vented, perhaps when methane hydrate – very unstable stuff – suddenly decomposes. The gaseous methane rises in the form of countless bubbles if it's predominate too low pressure and / or too high temperatures. This process is similar to the rise of carbon dioxide bubbles in a soda bottle that has been violently shaken. Since the average density of the gas–water mixture is significantly lower than that of water, the upwelling is either greatly reduced or completely put out of force: The plunging of a plane into a methane cloud must result in a crash as a result of a nose dive. Ships could sink without leaving a piece of wreckage not only due to a lack of upwelling, but also when gas rises like a fountain only at the bow or the stern. In the case of methane rising in smaller amounts, it is possible that vessels subside a certain level below their normal floating level. Also submarines subside and may subsequently impact the oceanic floor.

As the gas bubbles raise, the friction that is caused due to their contact with water, results in the formation of electric currents which lead to the generation of magnetic fields, because each electric field also leads to the generation of a magnetic field. This can lead to breakdowns of electrical and magnetic instruments, or do such phenomena have a deeper cause? Do such out gassing processes bear any relation to electrical discharges from the Earth's inner core?

This phenomenon caused by rising methane gas and also possibly electromagnetic discharge could be accountable for many unexplained occurrences in the famous Bermuda Triangle, whereby ships and planes have disappeared without trace. Huge methane deposits have been discovered exactly in this area because a destabilization of the gas hydrates can lead to a rapid rise "blow-out" of this methane. Only since 1971 did one start to take cognizance of methane hydrate, as its presence was discovered in the Black Sea to the amazement the experts.

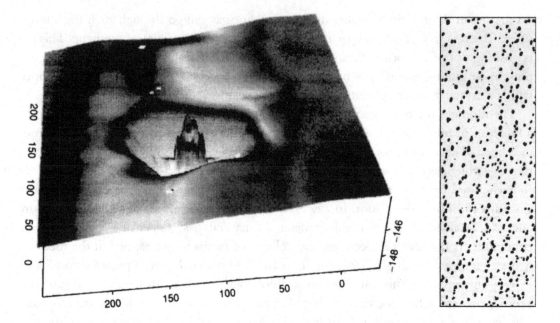

Fig. 3: Wet hole. Right picture: Off the Scottish coast, the floor of the North Sea is massively covered by craters (sea–bed pockmarks) in the area called "South Fladen". The picture on the right hand side shows an area of about three by nine kilometers littered with pockmarks (Judd / Hovland, 2007, p. 21). Left picture: Approximately 150 kilometers north of Aberdeen there is a crater called "Witch Hole" that is surrounded by "pockmarks". An intact steam boat from the early 1920th century is located in an upright position in the central area of the 120-meter-diameter crater (ibid., P. 373). Nothing points to an accident or collision. Alan Judd (1990) suspected that the boat must have been a victim of a sudden outburst of methane. Such eruptions are typical for this area.

By expeditions in the arctic Laptev Sea and off Pakistan's coastline – areas that both have boost abundant methane hydrates reserves – annular (ring-shaped) pockmarks were found on the oceanic floor: Craters with diameters of 20 to 30 meters that are apparently a result of gas eruptions (Kehse, 2000, p. 16). There are many others locations that are vastly dotted with such craters, some with an area extending over several hundred square miles. The area with the highest density of such craters has been ascertained to be in the Belfast Bay, Maine, that means off the U.S. northeast coast, with 160 sea bed pockmarks per square kilometer (Kelley et al., 1994). This research is published in the scientific journal *Geology* (vol. 22, 1994, p. 59), under the title: "Giant Sea-Bed Pockmarks: Evidence for Gas Escape from Belfast Bay, Maine "

It is alleged that there are conical depressions with a diameter of 350 meters and a depth of 35 meters in the seabed, that can be attributed to out-bursts of methane gas. In

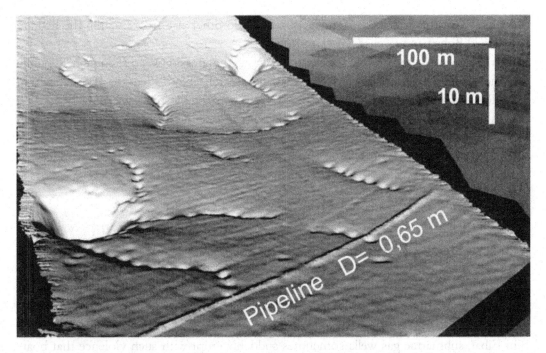

Fig. 4: Escape of gas. *Off the Norwegian coast in the European North Sea, there are rows of small craters (pockmarks) showing different directions, but either leading to or away from larger craters in a lined form.*

my opinion such geological formations around the world, even on land, have not yet been interpreted correctly, that is in the sense of outgassing processes.

As stated in the scientific journal *Geology* (vol. 23, 1995, p. 89), off the east American coast there are regular fountains of methane bubbles from the pockmarked ocean bed, that blow-up from a depth of 2167 meters up to a height of 320 meters, spreading over the vicinity.

In 1999, the German research vessel "Sonne" cruised off the coast of the Pacific coastline of the U.S.-state Oregon. Several clumps of white methane hydrate were discovered up to a size of a standard refrigerator, floating on the waves violent foaming getting smaller and smaller. "No one had previously observed that methane hydrate rises to the water surface and rapidly disintegrates, releasing methane directly into the atmosphere"(*GEO Magazine*, 04/2000). During dives on board the research submarine Alwin, "the scientists discovered several 15 centimeters wide vents in the gas hydrate, out of which beads of gas bubbles escaped. The geologist Dr. Gerhard Bohrmann (GEOMAR) suspects that the gas came from the area below the methane hydrate layer which is 140 meters thick, as shown by seismic measurements. "The methane must suddenly shoot through the hydrate layer – it would otherwise freeze within the hydrate layer", said Bohrmann (Kehse, 2000, p. 16).

In the course of the search for methane ice in the area between Georgian Republic and the peninsula Crimea aboard the Russian research ship, "Professor Logachev", in 2006 Russian and German scientists discovered remarkable holes in more than 1000 meters depth, from which sparkling fountains of gas ascended. "Such a one has not seen to this day, but it had also never been imagined that there something that looks like sludge, contains up to a fifth pure gas hydrate (...) and in addition there are traces of crude oil. A sensational discovery!"(German television *ARD Online*, "W wie Wissen", 5. Nov 2006).

Massive Stock of Methane Hydrate

"Submarine gas eruptions are known in the Caspian Sea, off the coastlines of Burma and Borneo, off the coastlines of Peru and in the Gulf of Paria between the northwest section of the island of Trinidad and eastern Venezuela. Off the coastlines of Baku at Bibi Eibat, submarine gas wells sometimes suddenly erupt with such violence that boats capsize when they came too close to the water swirls. If the sea is calm, such gas eruptions can be seen from afar. Off the coast lines of the southeast section of the island Trinidad, submarine gas explosions have been observed that eject water columns, accompanied by pitch and kerosene (Stutzer, 1931, p. 280). A huge deposit of methane hydrate that entails a 26 000 square kilometers large area has been discovered off the southeast coastlines of the USA. In this such a large quantity of carbon is contained that could cover the energy consumption of the entire U.S. for more than 100 years. Because there are even sometimes blocks of methane hydrate rising up to the surface, one can consequently speculate on the existence of sudden gas eruptions. The sudden release of large quantities of methane can perhaps also lead to environmental disasters.

Scientists make this accountable for massive greenhouse effects. According to the *geological time scale*, around 181 million years ago Ichthyosaurs, marine crocodiles, spiny sharks and Plesiosaurs were stacked in a mass grave at Eislingen in Baden-Wuerttemberg, Germany, almost like sardines three-dimensionally packed in a tin. The geoscientists at the University of Tuebingen presume the release of methane hydrate must have triggered this ecological catastrophe, whereby methane entered the atmosphere to set in motion a warming process of several degrees Celsius. Could this have been an early form of abrupt climate change?

In many places on the seabed of oceans, in the time period of the past 30 years found it were discovered large amounts of methane in the form of methane hydrate (MacDonald, 1997). Why is methane ice formed on the seabed? Since methane

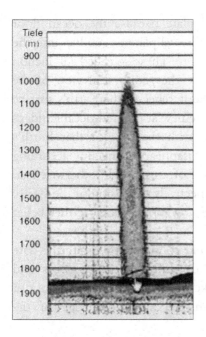

Tiefe (m)
900
1000
1100
1200
1300
1400
1500
1600
1700
1800
1900

Fig. 5: Gas fountain. *During the course of sonar measurements in the Black Sea, on board the research ship "Professor Vodyanitskiy" in May 2003 a 850 meter high water column was documented that had been caused by a sudden violent gas eruption (Egorov et al., 2003).*

lowers the freezing point of water, methane ice is formed even in areas where else is water in liquid form. The water thus freezes, because it's exposed to methane. Methane hydrate is for example formed at a temperature of less than seven degrees Celsius and a water pressure of about 50 atmospheres, which means below a water depth of 500 meters. Methane hydrate can therefore be formed anywhere, in areas with higher water depth levels, whereby the seabed has no warm water currents and methane flows out of the oceanic floor. In warmer areas (hot spots) or where the physical prerequisites for the formation of methane hydrate do not exist, the methane simply gushes out of the sea floor.

Between Greenland and Spitsbergen, an 1.3 kilometers wide and 50 kilometers long methane hydrate cushion has been discovered that is assumed to have a thickness ranging from 200 to 300 meters (Vogt, 1994). During exploratory drilling near Prudhoe Bay on the north coast of Alaska, large fields – each with at least eight seams – were discovered at a depth of 300 to 800 meters depth, which contain about 40 to 60 billion cubic meters of gas hydrate. A large gas hydrate has been discovered in Mallik, in the area of the Mackenzie River delta and the Beaufort Sea in Canada's Northwest Territories. Literally 80 to 90 percent of the pore spaces between the grains of sand and gravel are presumed to be filled with methane hydrate. Other explorations in Siberia and Alaska have reported concentrations of gas hydrate ranging from 50 to 80 percent. Marine deposits are larger, but it seems that they generally contain less than 20 percent. Surprisingly, gas hydrates in large amounts are increasingly being discovered, even in the Mediterranean Sea.

In deep sea regions, gas hydrates are partially found in areas with a gas–permeable seabed and in zones where cracks or fractures exist in the ocean floor – so called fracture zones – mostly alongside the steep continental slopes. According to the plate tectonics theory, methane hydrate accumulates there because in the belief of geophysics the oceanic floor shall drift like a roller conveyor, starting from the mid-*oceanic* ridges in direction toward the continental slope. Thereby methane ice moves with the oceanic floor and accumulates at the base of the continental plate because the ocean floor (respectively the lithosphere) should dive under the continental plate or subduct into the Earth's mantle within the region of a deep-sea trench (Fig. 6). These so called subduction zones are only based on a theoretically concept because the geophysical view of the world state a constant Earth's diameter respectively circumference of the Earth's surface and because "new" oceanic crust which is always originate new at the mid-oceanic ridges between the continental plates (continental drift) and with the proviso that Earth's diameter is constant, this "new" oceanic crust have to "disappear" stringently according to the plate tectonic hypothesis. Otherwise the new areas of crust would otherwise lead to a documented enlargement of the Earth's surface respectively expansion of the Earth. Therefore, subduction zones have been created *in the minds of scientists* because they belief that the Earth keep a constant diameter.

An expansion respectively growing of the Earth would consequently refute the core statement of the plate tectonics hypothesis in case of drifting oceanic floors and continent plates. If the Earth is expanding then gases like methane can easily gash out of fracture zones, which often exist in the thin ocean bed along the edge of the approximately 35 kilometers thick continental plate.

The hypothetical drifting scenario analog to the plate tectonic hypothesis is also inconsistent with the huge methane hydrate deposits which are deposited onto "normal" oceanic floors far from continental shelf areas. This fact is seldom mentioned and scarcely documented on maps which are showing methane deposits. Some researchers

Fig. 6: Subduction. *At the mid-ocean ridge (MR) between the continental plates, new oceanic floor is formed, such as has been ascertained through geodetic measurements. Therefore the distance between the continents will be longer. The whole originated oceanic floor respectively the lithosphere plate is presumed to drift from the mid-oceanic ridge (MR) like a roller conveyor and should dive under a continent plate (K) in direction to a subduction zone (S) – whereby the subducted plate areas equate to this ocean bed areas which has been originated at the mid-ocean ridge. If there is no subduction activity, this would lead to an expanding Earth, because, the new formation of "oceanic crust" results in a larger Earth surface. Consequently, if the Earth is growing the continents are relatively stationary, but nevertheless diverge from each other – without any drifting of continents. This signify: the plate tectonic theory is wrong.*

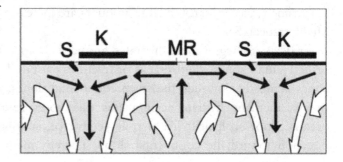

even assume that the largest methane hydrate deposits exist in deep sea basins and not along the continental shelf's (Klauda / Sandler, 2005). Estimates of the amount of methane are very uncertain, since there are hardly any observations of the several kilometers deep oceanic beds, nor sufficient drilling explorations. From the surfaces of planetary celestial body we know much more than from the depths of the Earth's oceans!

As soon as a sufficiently thick cushion of methane hydrate is formed in and/or on oceanic floor in the range of not too warm areas, then the methane – still upstreaming from below – accumulated under the already existing methane hydrate cushion to build a methane deposit thereunder. This continues until the dynamic pressure generated by the rising methane is less as than the pressure that results from the height of the water column and the weight of the soil layers. Below this stability zone or in the field of submarine volcanoes or hot zones (hot spots) the temperatures are mostly too high that the gas cannot be bound in the form of methane hydrate. Then methane exists in a liquid or gaseous state depending on the prevailing pressure and temperature conditions.

Methane hydrate deposits exist only on sea and oceanic floors? No! Already in 1976, methane was extracted from methane hydrates in the Siberian Messojacha Field with the help of an injection process. Another well-known example is the permafrost soil in the Arctic tundra, this stretches from Siberia through Alaska and Northern Canada up to Greenland. The geological thickness of permafrost depends on the air and soil temperature as well as on the features, especially the permeability of such ground floors. In some parts of Northern Siberia the permafrost can attain an extreme geological thickness of up to 1500 meters and even extend into the mid–latitudes (Nelson, 2003).

Since methane gas accumulates under the permanent frost ground, this may explain why the world's largest proven gas reserves are located in Russia. Globally almost one quarter of the world's land surface are permafrost zones, whereby the largest areas are located in the northern hemisphere (Zhang et al., 1999).

Moreover, the question is obvious why the ground floor can freeze to a depth of almost 1.5 kilometers, although the Earth is supposed to become warmer with increasing depth! However, methane exists in permafrost areas in form of methane hydrate, just as well as in and on the ocean floors. The amount of methane hydrates deposits in and under the permafrost is unknown. Estimates range from 7.5 to 400 billion tons of carbon (Gornitz / Fung, 1994). *Thus, in the permafrost is double or even a hundredfold of carbon saved, as is present in Earth's atmosphere.*

The sudden release of methane could be responsible for large scale natural disasters not only in the past but also up to today. Thus an "explosion" happened in 1908 in Siberia, known as the Tunguska event. Since this time many theories have been established thereof that range from the shattering of an UFO, over the aboveground bursting of a meteorite up to the impact of a black hole. The mystery is that no crater is present at the point of "must be epicenter", but starting from a certain point, 60 million

Fiction Subduction

Let us assume (contrary to reality) that the submerging tectonic plate is heavier than the material, into which it is supposed to be able to submerge. Taking into account the balance of all forces, the size of the applied force (N) at the plate end – tension (σ) multiplied by the area (d) – and additional the weight of the tectonic plate (G) must be greater than the opposing force (W) which is acting at the submerging end of the tectonic

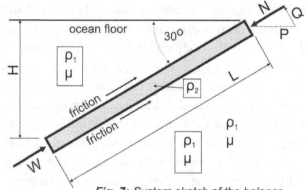

Fig. 7:-*System sketch of the balance of forces during subduction.*

plate plus the friction force (R), which is a result of the weight of the overlaying Earth's crust (H) plus the weight of the plate (L). The coefficient of friction (μ) is appointed to be constant for temperatures below 350 degrees Celsius (cf. Kirby / McCormick, 1982). This is the temperature that shall occur at the lowest point of the submerging tectonic plate (cf. Subarya, 2006, p. 50).

$N + G > W + R$ *(referring to 1 m width of the tectonic plate)*

$N + G = (6 \cdot D) + (\varrho_2 - \varrho_1) \cdot L \cdot D \cdot sin\ 30^o$

$(6 \cdot 10\ 000) + (32 - 28) \cdot 100\ 000 \cdot 10\ 000 \cdot 0,500$

$W + R = W + \varrho_1 \cdot H/2 \cdot L \cdot \mu + \varrho_2 \cdot D \cdot cos\ 30^o \cdot L \cdot \mu$

$0 + 28 \cdot (50\ 000\ /\ 2) + 32 \cdot 10\ 000 \cdot 0,866) \cdot 100\ 000 \cdot 0,6$

With these values the calculation can be reformulated to:

$(6 \cdot 10\ 000) + 2 \cdot 10^9 > 4,2 \cdot 10^{10} + 1,67 \cdot 10^{10}$

$6 > (4,2 \cdot 10^{10} + 1,67 \cdot 10^{10} - 2 \cdot 10^9)\ /\ 10\ 000$

$6 > 5,67 \cdot 10^6\ kN/m^2 = 5670\ N/mm^2\ > permitted\ 6\ up\ to\ 400\ N/mm^2$

Conclusion: A subduction is not possible, because the plate due to material-technical reasons tear before the force required for the subduction into the mantle is fully applied. Or in short: A subduction can not be attained neither through a pushing nor pulling force. Does the shear force (Q) respectively the redirection force in the wake of changing from horizontal to diving direction relating to force P already result to a breakage of the tectonic plate before pushing force P can be substantially applied? Every little "bulge" of the straight sketched plate leads to additional stress (bending moments) in the plate, through which tension forces generated that are non-absorbable from material- technical view point.

Remark: The at times advocated attempt by experts to lend credibility to the theory that the partial melting and the formation of moisture reduces the friction of thus, makes the subduction of the plate possible is wrong, because the frictional forces depend on the strength of the force and not by the size of the frictional area. This objection would be correct only if the entire plate would exhibit no friction area.

*Fig. 7a: **Tunguska Event. Left picture:** View from a helicopter: the southern swamp is the epicenter of the Tunguska explosion.. Photo by Vladimir Rubtsov (2009, p. 2), taken at June 30, 2008. **Right picture:** Trees were knocked down and burned over hundreds of square km by the Tunguska Event. Photograph from Kulik's 1927 expedition.*

trees are twisted outwards in a radial pattern over an area of 2000 square kilometers. Peculiar luminous effects were still observable at a distance almost up to 500 kilometers, in addition to pressure waves and sounds of thunder.

All attempts at explanation, which may include a physical–mechanical influence from the outside, do not explain why strange atmospheric luminous phenomena and minor earthquakes have occurred already in the days *before* the explosion,. If you are looking for a connection, then an interesting explanation of the Tunguska Event is a methane or natural gas outburst, as the astronomy professor Wolfgang Kundt (2005, p. 204 f.) himself explained me. For such an event a relatively small amount of about 0.1 billion tons of natural gas would be sufficient. In the underground of the Tunguska area are ample supplies of hydrocarbons, in particular methane, which is common in Siberia and the Artic Sea.

The pre-historical Storegga submarine landslide at the edge of Norway's continental shelf, beneath the European Norwegian Sea about 8000 years ago (according to geological time scale) is attributed to the progressive destabilization of gas hydrates. The required amount of methane is envisaged to have been a tenfold of that estimated for the Tunguska outbreak.

The Storegga landslide of the Norwegian costal lines is well known to the readership of the novel *The Swarm*. Frank Schätzing describes the formation of a massive tsunami that could threaten our culture today. However methane has been scientifically ransomed of being responsible for this landslide (Paull et al., 2007).

Fig. 8: Battle of Los Angeles. On 24 February 1942 several smaller points of light were sighted and in the early hours of next day there appeared a large luminous "Flying Object". This phenomena, interpreted as Japanese invaders, were taken under anti–aircraft flak from ground positions and through the air by airplanes. During this raid, however, no bombs dropped, and the air defence faded over 1440 magazines of ammunition, but it was not possible to damage one of the "attacker" or even to fired off. Damage occurred only by falling flak splitter and exploding grenades of the defender, not due to the "UFOs".

Frozen Mammoths

Perhaps another event should be investigated on the release of methane. As described in detail in my book *Mistake Earth Science* (Zillmer, 2001, p. 151 ff.), the mammoths died a cold death in present–day permafrost areas. According to reports, they were such in a manner quick–frozen in form of a snapshot that undigested grasses including buttercups, how was reported, have been discovered not only in the stomach but even also in the mouth. Mammoths are definitely inhabitants of a moderate climate, unable to live in frozen landscapes because they were equipped with a covering of long hairs. Therefore mammoths had a shock–frozen death partly next to rhinoceroses. But these were also discovered together with fleshy remains of horse (Ukraintseva, 1993), squirrel, rabbit, vole (Vereshchagin / Baryshnikov, 1982), lynx (Zimmermann / Tedford, 1976), bison (Anthony, 1949) and musk oxen (*Science News Letter*, vol. 55, 25. June 1949, p. 403). The ice with this animal remains must have been formed suddenly (see Figure 8)!

The meat of frozen mammoths is still so fresh today that that the locals use it to feed their huskies without any problem. Mammoth meat was even supposedly eaten by gold–diggers in Alaska during the Gold Rush. The ivory of the tusks is so fresh till today that the carving workshops in Asia are supplied with it.

With 16 percent the ice, in which the mammoths were frozen up, shows a substantially higher proportion of bubbles compared to glacial ice that only contains approximately six percent of bubbles. The structure of the "mammoth ice" is similar to that of granular hail, and some bubbles are interconnected or build a kind of chain in the form of a string of pearls. The ice also exhibits a yellowish color and often contains dirt and plant debris.

Fig. 9: Snapshot. During a journey through Tibet in 1848, whereby many travellers frozen to death, M. Huc (1852) reported that while crossing the Mouroui-Oussou river, which was frozen completely to the river bed, survivors discovered a herd of more than 50 wild oxen, which were suddenly frozen in swimming motion. The heads with the big horns still looked out of the formerly water and now ice surface while the oxen bodies were stuck in the block of ice completely. The transparent ice layer that stretched all the way down to the bottom of the river bed must have abruptly been formed to a big block of ice as a whole. This event is also reported by the founder of modern geology Charles Lyell (1872, p. 188) in the 11th Edition of his book "Principles of Geology". Image: Steve Daniels.

With the extinction of mammoths in today's permafrost regions of Siberia, the trunks of huge forests snapped off like matches and were drifted with the flooding of the rivers in the direction of the Arctic Ocean in the North. Did gas eruptions occur here? Methane oxidizes if air is supplied and out of methane there will originate carbon dioxide and water. This is an adiabatic effect *without any biological process.* Also the water entailed in the geological strata is released, and gushes out in the form of water fountains and mud volcanoes that can devastate whole landscapes with mud. The lungs of some mammoths were soiled and contained clay as well as sand particles.

These animals stuck in the mud which must have got frozen suddenly. It is assumed that the temperature must have fallen by perhaps to 175 degrees Fahrenheit (about minus 80 degrees Celsius) in order to freeze and preserve the stomach contents (Dillow, 1981).

How could the methane outbursts have led to such a sudden drastic temperature drop? Besides the shift of the Earth's axis and thus the climate zones, as I have discussed in my book *Mistake Earth Science,* Professor Karl-Heinz Jacob (TU Berlin) described to me, in the course of an information exchange, a hitherto disregarded aspect. He reported that in the course of "coal making", that is a extraction of coal with a pick hammer, this machine froze again and again due to the natural gas which was emitted from the geological layers and could no longer function until this pick hammer thawed again. He explained some more examples and referred to the *Joule-Thomson effect* that also could

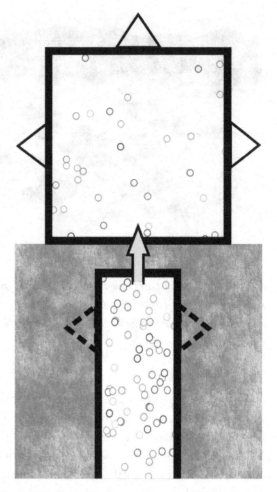

Fig. 10: Joule–Thomson effect. This adiabatic effect occurs when high–density gases by tension release through narrow fissures lead to a consequent pressure reduction. Concomitantly, the gas volume on the outflowing side increases, and at the die orifice occurs a strong cooling. In the Earth natural gas deposits in the lithosphere are also stored under high pressure. Through tectonic events such as earthquakes, fissures can arise, and the highly compressed gas can escape, whereby the Joule–Thomson effect is experienced. As natural gas with rich methane content is always compounded with groundwater (formation water) – which is also going to freeze in case of tension release because the Joule–Thomson effect occurs – we can explain physically the existence of sometimes deeply stretching "ice wedges" which can develop in "craters" of outflowing gas where the temperature is dropping down suddenly. Thinkable is also the formation of extremely deep stretching blocks of ice or ice layer. Such ice wedges or very deeply stretching permafrost formations cannot be built in case of cold weather in the atmosphere starting from the surface in direction to deep geological layers.

have also been responsible for the sudden freezing of the mammoths.

In thermodynamics, the Joule–Thomson effect (or Joule-Kelvin effect or Kelvin–Joule effect) describes the temperature change of a gas or liquid when it is forced through a valve or porous plug while kept insulated so that no thermic energy is exchanged with the environment. At room temperature, all gases except hydrogen, helium and neon are cooling down, which is accompanied with a contemporaneous expansion. This process is known as the Joule–Thomson process (compare Figure 10).

In fact, there are huge deposits of hydrocarbons in and under the permafrost regions in Siberia. As a result of the Joule–Thomson effect, gases that are under pressure *drastically cool down* as a result of the subsequent reduction in pressure. The volume occupied by the gas thereby increases dramatically. This is in effect, an adiabatic change of the state of gases, *without thermal energy being exchanged with their surroundings*. The Joule–

Fig. 11: Driftwood. At the mouth of the East Siberian Kolyma River, large logs have been documented that were washed up here as driftwood. Trees of this size do not grow today along this river. Also the coasts of the New Siberian Islands were covered by countless washed ashore tree remnants (Transehe, 1925). The trees snapped off by the blast waves of gas eruptions (see Tunguska Catastrophe), and the logs were buried in Siberia under the mud that gushed out of the erupting mud volcanoes as a result of gas eruptions. Tree remains were then transported northwards in the direction of the Arctic Ocean with the large streams, which eroded the just formed mud layers. This fossil wood is used for construction of dwellings by the locals these days.

Thomson effect may be experienced if for example, the high compressed natural gas existing in the lithosphere release tension while passing through narrow fissures. At the same time as a result of the pressure reduction, the volume of gas on the outflow side increases (in the atmosphere), since the average distance of the particles increases, and a strong cooling or freezing-up occurs at the die orifice.

A large–scale application of the Joule–Thomson effect is used for gas liquefaction. For example, if highly compressed gas at the target location of a natural gas pipeline is reduced to a technically usable gas pressure level through tension release, then the gate valves freeze and block the gas outflow. By the use of constant heating constructions, this freezing effect can be prevented.

The Joule–Thomson effect could therefore have been responsible for mammoths that were quick–frozen while feeding. May have been originated the whole permafrost soil in this manner? Mammoths grazed prior to this event in a flourishing landscape. The *Zoological Museum of St. Petersburg* has on display the fully–intact preserved Beresovka–Mammoth that had several kilograms of undigested plant remains found in the stomach, and blossom pollen in the long shaggy hairs. Calculations showed that the animal must have at least been quick–frozen at minus 65 degrees Celsius and must have been completely frozen within 30 minutes, so that for the stomach contents was not enough time to decompose.

Taking into account the vast areas in which animals were quick–frozen, a lot of methane must have gashed out, and almost abruptly. Is material fatigue of the ground layer responsible due to a high gas pressure from beneath or is it possible that catastrophic scenarios on the surface or in the atmosphere are parental accountable?

"On all the geological maps it can clearly be seen that on the island of Novaya Zemlya and the Taimyr Peninsula, there are rock deposits of the same old age. They geologically spread out to the northeast. The stratigraphic boundaries of their age structure, however, do not correspond with the geological spread. This is due to a steep horizontal displacement (fault)." (Drujanow, 1984, p. 96).

The geologist V. Ryabov reconstructed on the basis of paleo-techniques with the aid of aerial photographs an old river course and two fracture zones that run through the oil– and gaseous area and at which the crust has been shifted. The western part moved to the north, the eastern part to the south. Two gigantic geological slabs of the Earth's crust were thereby formed, whereby the edges are today's Western and Eastern Siberia. "The northern border of metal leading zone in Transbaikalia is shifted in comparison with such a zone in the area of Irkutsk. The displacement is approximately 500 kilometers to the south.

If one "moves" Eastern Siberia in the flow direction of the Yenisei River about 500 kilometers northwards, two excellently matching parts evolve. The Taimyr rock layers then form the continuation of the layers of Novaya Zemlya, which then correspond with respect to geological strike, and the metal leading zones of the Trans-Baikal and the Irkutsk regions unite again" (ibid., p. 95 f.).

Such displacement scenarios conflict with the plate tectonic hypothesis, but can be explained in terms of an expanding Earth. The Soviet geologist Vladimir Abramovich Drujanow (1984, page 69) writes: "Ultimately, the astronomers also render their support. With the help of atomic clocks they have found out that some of the time stations distributed in Europe, move to the east and others to the west. The simplest explanation for this is the expansion of the Earth."

Exclusively Organic?

Methane is the main component of natural gas and exists in large quantities on Earth. Estimates of the carbon amounts solely bounded in methane hydrates, ranges from 500 to 3000 billion tons (Buffet / Archer, 2004 and Milkov, 2004) or even 5000 to 12 000 billion tones ((Suess / Bohrmann, 2002). This is three thousand times the amount of carbon which is contained in the atmosphere. But, *such a large amount* of methane is still assumed to exist underneath the methane hydrate-deposits (Archer, 2005). Also the mine gas trapped in compact black coal deposits entails mainly methane.

Where does this incredibly large amount of hydrocarbons come from? According to the conventional-geophysical worldview, which constitute also the base of the climate policy, each molecule of methane respectively all methane on the Earth is ultimately attributed to a biological origin (Collett, 1994).

In order for methane to form all these organic materials, there must had been exist a tremendous mass of dead plants or animals remains, which were deposited. The carbon-rich pulp on the ocean floor must be covered quickly by sea mud and sand, so that the plant remains are not decomposed to carbon dioxide with the activity of aerobic (oxygen needed) bacteria. Instead of this conception, putrefactive bacteria, that require *no oxygen*, must play an active part to form methane – thus the conception may be correct.

No attempt whatsoever is made to take account of the fact that methane on Earth can also have an inorganic (abiogenic) origin, because the Earth is earlier supposed to have been a hot glowing ball! By such high temperatures hydrocarbons would decompose, and subsequently methane or ethane can only by formed organically through biological processes. Is this view correct?

It should be noted that methane oxidizes very rapidly in the atmosphere. Through this process, water and carbon dioxide is originated from methane. The half-life period of methane – that is the time in which the quantity is halved – is estimated at only 14 years in the Earth's atmosphere. Consequently, over long time periods of the geological history a fast regeneration of methane was required. This also counts for times when there was no life on Earth. From this perspective, is it possible to build large deposits of methane of solely biological origin? Since methane is also present on Mars, the question arises, why hydrocarbons can exist without any biological activity? Mars was supposedly also once a hot planet just likes the Earth! Let us therefore pose to have a look at our solar system and perhaps beyond this in outer space where methane is still present. Not so much time ago one would have laughed at this question. Because it was absolutely certain: Methane is of organic origin and exists only on Earth! But the opposite is true.

Methane in the Solar System

If hydrocarbons – such as methane or ethane – frequently exist on other celestial bodies, we would argue that the hydrocarbons there are formed in a completely different manner as those on Earth, since neither complex biological nor photosynthesis processes exist. If a simple life may exist in the form of microbes, these biogenic processes are too small to produce large quantities of hydrocarbons. Consequently, the large supplies of hydrocarbon on other celestial bodies originate from an unknown source material, which can produce the carbon that contains in the atmosphere of Mars in the form of carbon dioxide. Does such inorganic (abiogenic) process take place also in the Earth?

The former investigators knew only organic starting material and the biological process of photosynthesis. They ignored that carbonaceous chondrites (Wilkening, 1978) – a special form of the stony meteorites – contains up to three percent of carbon, which is largely non-oxidized and consist of up to 20 percent water. These non-oxidized carbonaceous materials of such meteorites – which supposedly exist since the formation of the solar system – requires a low-temperature condensate that could have at *no time* been exposed to higher temperatures until today! Only such carbon, kept under cool condition, is received in non-oxidized condition. Can we identify carbonaceous chondrites material as a component of the Earth?

Isotope data's for noble gases like neon, argon and xenon trapped in the crust of the Earth show a closer analogy to the noble gases existing in carbonaceous chondrites than in any other kind of meteorite materials. Entered carbon, as some scientists mean, the primeval Earth with meteorites? Or was, seemingly heretically asked, the Earth at low temperatures in the form of a low-temperature condensate such as carbonaceous chondrites? In this case, non-oxidized carbon was contained in the Earth from the beginning until today like in carbonaceous chondrites and other comets. But then, consequently, the interior of the Earth was and is not hot or was glowing in the past. The Sun was always relatively cold like carbonaceous chondrites and comets.

Since it is believed conventionally that the supposedly earlier hot Earth must have oxidized all carbon supplies on Earth, hydrocarbons such as methane can only be developed subsequently in the cooler crust and not inside the Earth. For this case, hydrocarbons may be only of organic (biologically) origin. Therefore, it was a surprise to the scientific community, that the intensive satellite research proved that methane was present in the atmospheres of Mars, Jupiter, Saturn, Uranus and Neptune, but also on Jupiter's moon Titan or the dwarf planet Pluto.

Actually, there was announced the discovery of methane on Pluto's surface and in the atmosphere. There are large areas covered with methane ice, indeed, it is even given "boulders" of methane. One suspects a climate change caused by methane, because the temperatures near the surface went to stay 40 degrees lower than in the atmosphere of Pluto (ESO press release of 2 March 2009). The dwarf planet Pluto was fumigating!

Spectroscopic analysis of Pluto's surface shows also that here is nitrogen ice, with traces of methane and carbon monoxide (*Science*, 2007, vol. 261, p. 745–748) and in October 2006, the *NASA/Ames Research Center* announced the spectroscopic discovery of ethane; which is second-largest component of natural gas. This hydrocarbon should only occur biologically on the Earth. Why not also abiogenic?

More recent measurements by Stephen Tegler (*Northern Arizona University Flagstaff*) in 2008 showed that the surface of the dwarf planet Eris, which is 100 kilometres larger than Pluto, is covered of frozen methane. One doesn't exclude gas eruption from the interior of Eris. Such a scenario is in contrast to our astronomical worldview; because this dwarf planet reside in the ice cold space beyond Pluto and orbits the Sun in only 560 years.

On Earth, the methane content in the atmosphere is only 1.75 ppm (parts) of one million – according to the present conviction. On the other hand, the methane content in the atmosphere of Jupiter (3000 ± 1000 ppm) and Saturn (4500 ± 2000 ppm) is more than one thousand times as high.

But according to an article published in the journal "Science" in 2009, a study of Italian geologists emanate more hydrocarbons – such as ethane and propane – from gas deposits in the Earth's crust than thought up to now, after the concentration of gases in the atmosphere have been investigated at 238 localities in the world. This investigation shows that in the record of climate models, large quantities of hydrocarbons, so-called greenhouse gases, are not included. The ethane as well as propane content in the atmosphere was identified in each case with 15 instead of the 10 million tonnes previously considered. It is a principle error in climate models, because the additional emitted hydrocarbons are indeed not caused by plants, bacteria and human causes, but come from a previous source in the Earth which was not taken into account. Additional degassing of methane was generally not included in climate models because this hydrocarbon is largely oxidized already in the crust, whereby water and carbon dioxide (abiogenic) arises, and only a part of methane is degassing in the atmosphere. The proportion of greenhouse gases is also not considered by climate scientists. The Italian geologists confirmed: "Our studies refute the conventional view that a purely geologic emissions of hydrocarbons, will only have a negligible exert influence on the atmosphere" (Etiope / Ciccioli, 2009).

The question of the origin of methane on other celestial bodies is of fundamental importance, as there hydrocarbons must be originated inorganic (abiogenic) while such hydrocarbons on Earth are to be exclusively results of biological (biogenic) processes. Therefore, there is absolutely no explanation for the existence of methane on other planets. In concrete terms: large quantities of hydrocarbons occurring on other planets contradict our conventional worldview in a fundamental way. Therefore, the conventional theories about the origin of our solar system and the planets must be incorrect. The formation of methane in the interior of planets is not compatible with a presentation of formerly "hot" planets or other celestial bodies! High temperatures would have decomposed the carbon at an early stage, and there is no thinkable worldview to build hydrocarbons like methane during a period in the aftermath of the hot phase. Only on Earth there should be an exception mechanism: namely, the photosynthesis! Therefore, all hydrocarbons are seen on the Earth should be originated as biogenic. Is there carbon, for example, on and in the Mars since its formation, this planet has always consisted of a low–temperature condensate all the geological time until today. Similarly by the same developmental history, also on Earth, deep and not very high temperatures should have prevailed. Therefore the Earth has always existed as a low-temperature condensate if there is no oxidized carbon since the beginning of our world in the depths of the Earth. Are there any carbon examples, existing on Mars since its formation, then this planet, as well as comets and asteroids, has also always exist in form of a low–

temperature condensate. In case of a similarly history of developmental of celestial bodies, also on Earth, deep and not very high temperatures should have prevailed. Does our Earth always exist in form of a low–temperature condensate?

Let us therefore consider first, in which form carbon exists. In the atmosphere of the outer planets hydrocarbon is *not oxidized*. In contrast, the inner planets exhibits carbon mainly in *oxidized* form. If we assume an uniform origin of the inner and outer planet since the forming of the solar system, then chemical processes are responsible for the oxidation of carbon at the inner planets; after the building of proto–planetary bodies.

Indeed, in the atmospheres of the inner planet's oxygen is produced without biological processes, because the sunlight splits water molecules into oxygen and hydrogen. The lightweight hydrogen escapes into space, leaving oxygen in the atmosphere, which can initiate oxidation processes. This abiogenic process to form oxygen is mostly not mentioned and the Earth's oxygen production is fully attributed to the photosynthesis process and is therefore considered to be biologically. However, "the oxygen in the Earth's atmosphere today is of abiogenic origin" wrote my friend, the renowned German astrophysicist Professor Wolfgang Kundt (2005, p. 204).

Assuming that the primordial Earth was not hot, then there may have existed water immediate since forming of the Earth like on comets and asteroids. In a gas cloud which was collapsing to build a star, astronomers discovered so much water vapour that was five times more than all the oceans on the Earth. For the first time, Researchers at the *University of Rochester* showed on the basis of recordings, how water is transported to regions where planets may be formed thereinafter (Watson, 2007). Since the matter in the universe consists of 90 percent of hydrogen (ionized in the plasma state), plenty of water may well occur, because plenty of free oxygen exist in our solar system – without plants and photosynthesis..

On Mars, rocks on the surface are in oxidized state. The red planet owes its colour to the oxidation of iron to iron oxide, respectively the chemical compound of oxygen and iron. Mars is therefore a *rusty* planet. In January 2004 U.S. probes which were landed on Mars discovered two minerals on the ground which only arises with the participation of water: goethite and jarosite. In both minerals on the one hand, oxygen is chemically bound and on the other hand, are both regarded as evidence that water was previously available on the Martian surface.

Today, the atmosphere of Mars – together with traces of carbon monoxide and water vapour – consists of only 0.14 percent of oxygen, but 95.3 percent of carbon dioxide. The two eye-scatching ice caps at the poles consist mainly of frozen carbon dioxide and a small amount of water ice. Why is there so much carbon dioxide on Mars? In the atmosphere of the Earth is as low as 0.038 percent by volume. This question can be answered, if there is methane in the interior of Mars, which diffuses partially into the upper rock strata, and in the end from there into the atmosphere.

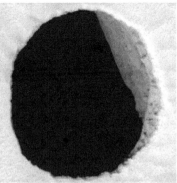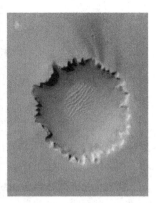

Fig. 12: Vents on Mars. On 5 May 2007 the NASA Mars space probe Reconnaissance Orbit photographed a vent with a diameter of just less than 160 meters and a depth of at least 78 meters (marked by an arrow in left photo – enlargement of this photo in the middle). It will be a "lava tube", although there is no lava. Escaped here as well as gas from the former Victoria Crater (image on the right page), which should constitute an impact crater (Fig. 13), despite the shallow margins and the lack of a central peak and also of ejected material? The Victoria Crater is probably just a small mud volcano, the vent was closed by fine-grained sediments, which were turned into mud by the presence of water – originated with the oxidation of methane. In the middle of the crater are still sand dunes to see in the right picture. Outside of the crater there are still visible "carbon black plumes" which occur with the blowing out of the crater in case of incomplete combustion of hydrocarbons.. Photos © NASA.

The degassing methane is mostly already converted in water and carbon dioxide with the oxidation process and this take place as is mostly the case already in the Martian crust as soon as it comes into contact with oxygen. This oxygen will be partly consumed by the oxidation of hydrocarbons and is partly bound chemically in rocks: This is the reason why there is low percentage of oxygen in the Martian atmosphere. Are carbonaceous gases degassing from the interior of the planet? In this case more and more carbon dioxide is going to be accumulated in the atmosphere.

In this way, respectively with the oxidation of methane on Mars, on the one hand we are able to explain the enigmatic partial water resources and on the other hand, the extremely high content of carbon dioxide discharged in the atmosphere. Therefore, water in liquid form should be found *below* the surface and also emerge through fractures into the Martian crust, because methane as a result of different geological structures is not distributed evenly, but should form a pattern of increased concentrations. Water, methane or carbon dioxide can even shoot into the Martian atmosphere like a pulsating spring.

In fact, it has discovered specific, among others ring-like pattern and there is evidence of huge Martian geysers, which skyrocketed fountains of carbonated water up to heights

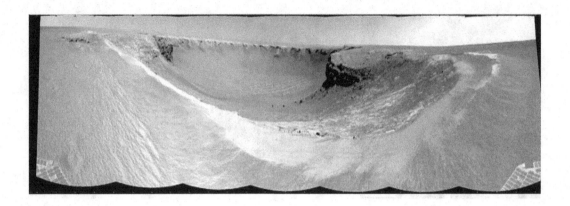

of several miles. The muddy water, so the researchers suspect, rained back on the ground again at a distance of several kilometres from the rising. However, not all the water falls back on the floor in liquid form: Especially in the boundary areas of the fountains, the water should freeze rather quickly at a temperature of minus 70 degrees Celsius to fall back on the Martian surface in the form of hail. Water may have been ejected very constantly over a time period of one or two months (Shiga, 2008). Such a scenario was able to shock–freeze the mammoths in Siberia in areas with abundant methane deposits and mud volcano eruptions in today's permafrost.

In principle, the described Martian geysers are the same phenomena as the enigmatic mud volcanoes on Earth. The several hundred kilometers long rift and crack systems *Cerberus Fossae* and *Mangala Fossa* on Mars are starting points for wide channels. For such territories is already assumed that there were flowing large water masses some times in

Fig. 13: Impact craters. These have, – in contrast to the Victoria crater – a raised rim, a central peak and ejected crust material around the crater. The rock in the crater is partially melted and in the underground crashed. Compare this with Figure 12

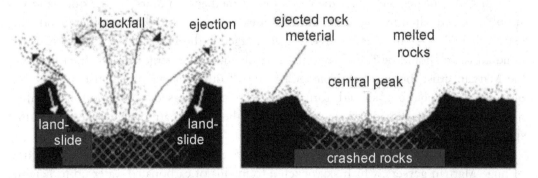

the past. It should be to find holes in the Martian soil, from which gas e scapes, similar to the bottom of terrestrial oceans. Such tube–shaped openings are now photographed as to see on Figure 12.

The repeated discoveries of methane on Mars have not let to a formal debate going on. Perhaps scientists do not want to speculate on volcano activities on the one hand or actually about life on Mars on the other hand. The real problem is that the origin of methane – or more generally of hydrocarbons – without any biological origin cannot be officially explained. Therefore, the researchers are looking for life on Mars and now it is suspects that micro-organisms are living under the surface of Mars – even after this had been disputed since the landing of the Viking probes. These organisms are said to have produced methane for three methane clouds which had been recently discovered. The largest of these sources will emit at least 36 kilograms of methane per minute (see Mumma et al., 2009).

Now it is believed that micro-organisms to be responsible for the production of methane; because the possibility of abiogenic origin is absolutely excluded.

Roughly 100 years ago, as the theory of the biological origins of hydrocarbons was developed, the supposedly secure knowledge was very different from those of today. At that time, one still believe that the Earth formed itself as a globe in a hot fluid state and subsequently very slowly cooled down, in order to form a solid crust as a result. Since methane and other hydrocarbons would not have survived at the high temperatures, how such should exist at 10 or 15 km depth, an origin of hydrocarbons in the Earth's crust and an adjusting degassing process temperature was simply unthinkable due to the assumed high temperatures. However, higher temperatures in the depths of the Earth are accompanied by a high pressure, and it has only been recently recognized that large hydrocarbon molecules are stabilized by pressure. For their origin, a depth of 150 to 300 kilometers is quite possible (Chekaliuk, 1976). Maybe the lowest areas in which methane exists in the Earth may be at a depth of about 600 kilometers (Gold, 1999, p. 50).

Methane in this depth refutes the current world view. To deny the existence of hydrocarbons in the upper mantle calculations have been made. This, however, requires a chemical balance and is correct only under these conditions. The calculations are not meaningful, because one conventionally assumes only small amounts of methane as well of other hydrocarbons If one considers only the unimaginable quantities of petroleum, natural gas and methane hydrates on the Earth, then very large and not just small amounts of gas – rising from the depths (like methane) – are to be taken into account.

It necessarily follows that – contrary to the conventional view – no chemical equilibrium in the Earth's mantle and the crust may be present. Therefore, hydrocarbons may also have been preserved in large quantities (Aculus / Delano, 1980).

Only gases (in liquid state) in large quantities are able to set mechanical processes in motion that cause fractures in bedrocks or sediments and thus to start a degassing process through solid rock. *Small* amounts of gas in the sense of a chemical equilibrium are not able to do so. On the other hand large quantities of gas in such fractures at this depth come only into contact with relatively *limited* amounts of rock surfaces. The oxygen, available due to diffusion, is limited, so only a *limited* amount of rising hydrocarbons (such as methane or ethane) can be oxidized. As soon as the oxygen is used up, further hydrocarbons degassing from the depths of the Earth through the already opened are able to degas through this rock stratums *without* being oxidized. The oxidation then only takes place at a higher stratum, where oxygen is still present. In this case hydrocarbons are rising higher and higher through the crust without any oxidization.

The chemical disequilibrium of a large quantity of methane with relatively little rock is not comparable to the case of a chemical equilibrium of a large amount of rocks with little methane: If methane and other hydrocarbons, to a large extent survive in the temperature/pressure areas at depths of several hundred kilometers and if hydrocarbon liquids are present in sufficient quantities these are able to break open cracks through solid stratums and thereby remaining undestroyed, the origin of methane and petroleum in very deep strata's of the Earth must then be taken into account as a serious possibility (Gold, 1987, p. 15 ff.).

This idea is, however, literally dogmatically blocked by the theory of the hot Earth. Let us therefore take a look at the cold Mars to break this blockade. Let us now assume that methane is degassing continuously, according to the scenario outlined before, from the depths of our sister planet, then the carbon dioxide is consequently a result of the partial oxidation of hydrocarbons and is thus not the direct (primary) product of a previously *unknown* material source, whose existence is actually unknown and whose existence has yet to be searched. In other words, in the established world view there is no source material for the carbon dioxide – however also not for the hydrocarbons – on Mars.

The Martian rocks on the surface feature a bubble-like structure. They are similar in their composition to terrestrial basalts. Partial melting of the rock tends to bring more oxidized rock to the surface because it is less dense and thus able to ascend to the surface more easily – similar to the salt in the Earth's crust (description in my book: *Mistake Earth Science*). One knows the old phenomenon that new pieces of stone continuously "grow out" of arable land on Earth. This is the reason why the inner planets exhibit strongly oxidized conditions that we find today, and there are no conclusions possible, from which one can discern how large the supply of carbon was, because carbon would have had chemically combined with the existing oxygen chemically and produced carbon dioxide. The biological processes on Earth only reinforce this process, *but do not originate it.* This idea crashes our conventional world view.

The polar caps of Mars show spiral cuts. There origin has yet not been explained. At the south pole of Mars an impressive spectacle takes place in spring every year: when the ice layers on the polar cap heat up after the cold winter, everywhere geysers of

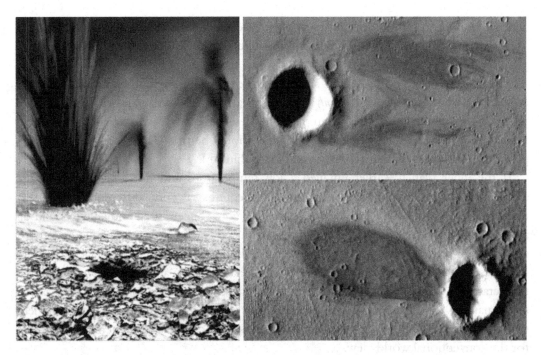

Fig. 14: Mars geysers. Picture Left: *NASA artist's impression of considered possible (carbon dioxide) gas jets at the south pole of Mars (Image: Ron Miller, Arizona State University).* **Pictures right:** *Several images taken by the Mars Odyssey spacecraft show circular craters that have a graded coloration caused by gas eruptions on the Martian surface. These can hardly be regarded as impact craters (compare Fig. 13), since no ejection material exists around the craters. Even in the area of the dark discoloration smaller circular degassing craters (pockmarks) can be seen.*

carbon dioxide and dark sand–dust start to shoot up. This dust then finally falls to the ground, where it forms flashy dark spots and fan-shaped structures. This is the conclusion made by astronomers around Philip Christensen, based on new data from the spacecraft Mars Odyssey. But perhaps, is a different process decisive as the warming of the surface by solar radiation? Are this geysers little methane and / or carbon dioxide volcanoes, which testify degassing processes?

If there are carbon dioxide (and methane?) emitting geysers on Mars, then this means that the material that forms the upper strata on Mars is not well mixed and contains large amount of volatile substances. In a well-mixed Martian material with a small proportion of volatile substances, these volatile substances would remain locked forever and could never get to the surface. There must be a continuous increase of volatile materials, in order for the geysers to receive supplies. The quantity of these substances must be large enough in order to rupture rock layers and for gases, as a result of the overpressure in the Martian crust, to shoot up like a fountain. Small quantities of gases would not be in the position to create and supply geysers.

If Mars was earlier not cold, but supposedly hot like the Earth, then a *high temperature* condensate would have been the *only* basic material. This must have inevitably contained much less volatile substances than can be detected today, whereby a subsequent formation would be excluded if our world view should be true. Therefore, was the Earth previously hot, gas eejecting geysers would be unthinkable.

If the inner planets earlier entailed a low-temperature and not a high-temperature condensate, then a later degassing process requires a very differential unequally distribution. Only if liquids and / or gases were highly concentrated and able to build up a correspondingly high pressure, rock layers could be cracked. Gases ejecting geysers or only springs with gushing water are therefore only possible if fracture zones were built. Furthermore, it is also clear to understand that relatively low and *on no account* very high temperatures are necessary, otherwise methane would have decomposed, which is the basic material to build carbon dioxide and water in case of oxidization. This statement must be underlined: Although it sounds harmless, this conclusion is mentally explosive for the conventional world view.

At the autumn 2006 meeting of the *American Geophysical Union* (AGU), Tobias Owen (University of Hawaii) and his colleagues reported, that some larger objects in the Kuiper Belt, which is assumed to exist behind Neptune's orbit, appear to possess a "methane factory". Methane can occur in all three aggregate phases – solid, liquid or gaseous – how they altogether exist on Saturn's moon Titan. There are supposedly even clouds of methane and other hydrocarbons like ethane, from which it rains to build "methane seas". Nevertheless, there is no greenhouse effect on Titan! Only because this Saturn moon is too far away from the Sun and therefore too cold?

Let's have a closer look at Saturn moon Titan, which we particularly know very well as a result of the landing of the Huygens probe on 14 January 2005, and measurements by Cassini. Recordings by the Cassini spacecraft on 26 October 2004 brought evidence of a circular structure of about 30 kilometers in diameter, possibly a volcanic crater, called a caldera.

This volcano would however, not emit glowing lava, but a mixture of water and methane. Thus there is a subsequent delivery of fresh methane. According to the findings of the Huygens probe, constantly, at least periodically, methane is produced in large quantities on Titan. This, one can read in the ESA News, 30 November 2005. Such a process must continue over long periods, otherwise methane would have decomposed due to the UV radiation from the Sun within a time period of few years. How is this methane originated according to the opinion of the researchers? Of course inorganic (abiogenic): Molecular compounds from the interior are supposedly degassed and converted into molecular nitrogen, whereby carbon monoxide with hydrogen is reduced to form hydrocarbons. Methane is therefore accordingly formed without biological processes in the interior of Titan. Does this process also occur on the Earth like on other planetary bodies? But then, there must be nitrogen, carbon monoxide and hydrogen in the depths of the Earth.

Inorganic Origin

The theory, whereby hydrocarbons are formed inorganically on Earth, is not new, but was developed already in the 19th Century in Russia. What followed was a good deal of thoughts down to the present day, (such as Kropotkin, 1985) to the inorganic source of hydrocarbons. The passed away Austrian professor Thomas Gold, who can be counted among the most important scientists, discussed the originating of carbon material as a consequence of a flow of hydrocarbon fluids from deep sources of the Earth and was convinced of the abiogenic origin of oil and gas (Gold, 1999).

Through a Swedish government-funded drilling in 1983, Gold was able to demonstrate that methane, tar and oils are upward flowing (migrate) through the Swedish granitic basement. This nevertheless does not come officially into question as a source region for hydrocarbons. Because there are no thick sediments, the origins of methane (natural gas) and petroleum must be located at even greater depths.

In deep drilling in the Kola Peninsula in the far north of the European part of Russia, was found in rock – with a similar crystalline structure as in Sweden – crude oil in a depths of 11 000 meter. However, if hydrocarbons are of biological origin, they should mostly to be found only in fossils containing sediments but not in granitic basement, whose fissures and cracks can only be infiltrated by hydrocarbons in small quantities.

Methane appears to exist deep below the Earth's crust and should therefore also be originated there. But also carbon dioxide comes from the depths of the Earth, as volcanic eruptions vividly demonstrate before our eyes. "If carbon dioxide as well as methane exist at these deep levels – and there is evidence for this – then it may also be that carbon dioxide resulted from the oxidation of some of the methane, rather than being a primary product from a source material" (Gold, 1987, p. 36).

In other words, if methane and other hydrocarbons migrate upwards from below, these could become the main source of the deposits of non-oxidized carbon in the soil. In this case, methane is the source for carbon dioxide, whereby with this oxidization process also is originated water.

In contrast to this thought, the scientific dogma asserts that there is a carbon cycle and all hydrocarbons on Earth are derived from organic materials, created through photosynthesis. Based on these considerations, it begs for the question as to where the huge amounts of carbon in the form of coal or oil come from. The earlier Earth shall have possessed a thick atmosphere of carbon dioxide, which reacted with calcium oxide to form the masses of carbonate rocks. These supposedly sunk into the depths at the joints between the plates (subduction zones), where the carbonate molecules decompose due to the prevailing heat. Carbon dioxide is released again, and spewed into the atmosphere by volcanic eruptions. Thus, a cycle supposedly arises that entails a consistent deposition of carbonates and release of carbon dioxide. If we are want to explain the gigantic carbon abundance in the upper soil layer, the original carbon dioxide atmosphere must have been very tremendous.

From the amount of known carbonates one can project that the mass of carbon dioxide in the primordial atmosphere was about eight times greater than the total mass of the current Venus atmosphere. In contrast, there is enough evidence that the Earth did not earlier possess such an amount of gaseous material. One of them is the low proportion of noble gases in the present atmosphere.

The theory of a dense carbon dioxide atmosphere defaults also in terms of the chronology of the geological formation of carbonate rocks. If the biogenic theory should be correct, the rate of the formation of carbonate rocks would have to reduce with decreasing amounts of carbon dioxide content in the atmosphere in the course of the Earth's history. However, exactly the opposite has been proved. The sediments have in the past two billion years of geological time, that is the period over which realistic conclusions in case of sediment discharge can be drawn at all, an overall steady increase and not the expected reduction of oxidized as well as non-oxidized carbon can be realized. In these 2000 million years the carbon supply in the near-surface deposits would have had to be renewed nearly 740-times, since it requires almost three million years to replace the whole today existing amount of carbon dioxide in the atmosphere and the oceans, because the present content of carbon in this pool represents only about one part in 740 of the known deposited amounts (Gold, 1999, p. 63)?

In other words, carbon dioxide must be continually newly originated and for this process, the degassing of volatile components from the Earth's interior can be made responsible. Let us therefore summarize this process briefly once again. The upward migrating carbon fluids, in particular methane, are oxidized on their way to the surface of the Earth. The ratio of methane to carbon dioxide depends on the available oxygen in the rock as well as the prevailing pressure and temperature conditions. That proportion of methane that reach the atmosphere without prior oxidation is relatively quickly converted in the oxygen-rich atmosphere into carbon dioxide *and* water by a half-life of only 14 years: The reaction of a methane molecule with two oxygen molecules result in the formation of two water molecules and a carbon dioxide molecule. In this way new carbon dioxide is always *abiogenic* originated all the time. On the one hand this enriches the Earth's atmosphere and on the other hand the oceans with carbon dioxide. With an increasingly warmer climate, carbon dioxide is also released from the oceans into the atmosphere without any human activities.

With volcanic eruptions ejected carbon dioxide and methane have been well studied with respect to the degassed amounts. But the large quantities of abiogenic gases that flow up from non-vulcanized soils have not even been measured to get an amount of such emissions in the atmosphere. Accordingly, climate activists only take methane into account for the settings of their computer simulations, that is biologically produced such as those from cattle and rice fields. It is official believed that 70 percent of the methane discharge is caused by humans. An alleged increase in methane levels in the atmosphere of about two-fold since the 19th Century is supposedly attributed to agricultural cultivation and intensive livestock farming, whereby the volcanoes are considered to be only a negligible source to emit methane gas. Actually there are, as previously described, huge amounts of (abiogenic) methane, other hydrocarbon gases and carbon dioxide *not* taken into account.

Methane Volcanoes and the Greenhouse Climate

60 million years ago (according to the geological time scale), quadruped, amphibian living pantodonts with a weight of about 400 kg existed on the arctic Svalbard archipelago. This heat-loving prehistoric mammals looked like a mixture of hippopotamus and tapir. 55 million years ago it became even hotter in the Arctic Ocean. At that time, crocodiles and turtles were living in the "Arctic". The average temperature rose to twenty degrees Celsius. At this time existed extreme greenhouse conditions on the Earth – including in the Arctic, said expedition manager Jan Backmann (*University of Stockholm*), according to a *dpa-* report of 6 September 2004:

"Based on our preliminary findings we have now to recalculate the early history of the Arctic basin. Actually the climate at that time was more changeable as we have thought before."

In the rocks of this time, geologists today find evidence of a rapid increase in methane content of the air. Thereafter subtropical marine algae swam in the around 20 degrees Celsius warm Arctic Ocean. Therese first results of the international North Pole expedition ACEX (Arctic Coring Expedition) were presented by the *Research Centre Ozeanraender* at the Bremen University in 2004. According to the opinion of the researchers a violent heat occurred in the Northern polar region, which triggered a mass extinction of marine life. Due to the high global temperatures the Arctic was then naturally free of ice, as well as Greenland and Antarctica as described in my previously books (Zillmer, 2004). In the history of the Earth, there have repeatedly been rapid climate changes, without any human influence.

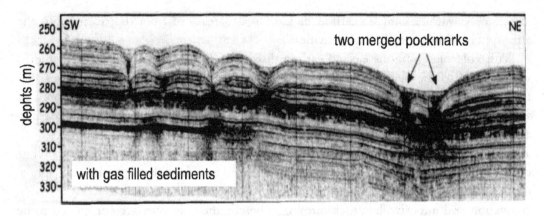

Fig. 15: Periodically. *Pockmarks over gas-filled sediments testify the existence of periodically occurring blowouts. A large crater was thereby formed from two smaller pockmarks (Cifci et al., 2003). In this section of the Black Sea the flow path of the gases through the sediments are clearly identify.*

There are vast quantities of methane gas under the North Sea. Holes have been discovered in this sea ground, which are supposed to be meteorite craters according to the general conviction. But in the majority, however, this should be holes in the sea ground, from which methane gas leaks. Due to the shallow depth of the water in the North Sea no methane hydrates can be formed here. From deeper layers upward migrating methane is converted by oxygen into carbon dioxide and water (oxidation process) already in the Earth crust respectively in the seafloor.

But, in case methane can migrate upward as such are oxidized, this reaction takes place in the sea water or when the saturation point is exceeded, at least only in the atmosphere. Whether and how much methane is released in the North Sea, directly in the form of gas or indirectly as carbon dioxide, has not been subject to official measurements or observations and is therefore unknown.

The mass release of natural gas (methane) is therefore directly emitted from "Methane Volcanoes"., which are not to be discussed in the literature. These gas volcanoes can man-made brought to outbreak, for example with the drilling into an over pressurized gas bubble. Something similar happened when the oil company *Mobil North Sea Limited* drilled for oil in the North Sea in 1990. The pressure of the methane bubble erupted spontaneously on 21 November 1990, this led to an explosion that destructed the oil rig. The drilling activities were giving up after this accident (see photo 12).

Since that time, there have been huge quantities of gas that have escaped in different strengths from holes that broke open after a depression of the sea ground. Researchers from the IFM-GEOMAR (*Leibniz Institute of Marine Sciences* at the *University of Kiel*)

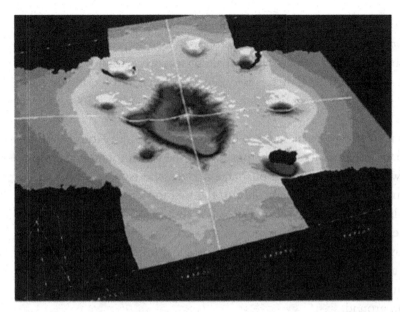

visited the original drill hole for the first time in 2007 with a research submarine. Currently, about 1000 liters of gas escape per second from the drill hole. "The unique thing here is," says the expedition leader, Olaf Pfannkuche", that free carbon dioxide and methane reached the water surface" (*www.welt-online.de*, May 14, 2007), and to be exact this is approximately one third of the total amount of gas.

Certainly during the explosion in 1990 and shortly thereafter, a significantly large amount of gas was blown into the atmosphere. Let's us assume that the gas currently escapes at 1,000 liters per second or 333 liters, which is in any case the amount that is blown into the Earth's atmosphere, and let us set this emission rate as constant for the time since 1990. Then arithmetically by the end of March 2009 approximately 193 billion liters of methane would have been blown into the atmosphere, in addition to the carbon dioxide that is originated when the gases migrated upward through the Earth's crust or in the waters of the North Sea. In addition, there is originated water as a result of the oxidation of methane. Therefore, in this way additional sea water is created in the sea. This all involves only *one* well-known, relatively small bore hole in the seafloor of the North Sea. There are also others, much large sources of methane.

The huge gas deposits beneath the North Sea alone bear a potential risk of a greater methane explosion. Perhaps this is the reason why the isle Iceland recently subsided dramatically. In 1908, Professor John W. Walther wrote in his book "History of the Earth and life":

"Currently, as we to be informed of Nansen keen boat trip (with his research ship

Fram), the largest part of the Arctic sea is deep-sea floor, but numerous shells of *Yoldia artica* . . ., *Cyrtodaria siliqua* and numerous pieces of otoliths – a structure of sensory cells situated in the inner ear – of fishes living only in shallow seas were found on deep-sea grounds at deep depths of 1000 till 2500 meters between Jan Mayen and Iceland, because this part of the Arctic Ocean has been subsided in recent years around 2000 meters" (Walther, 1908, p. 516).

He also points out that it had to come hand in hand a substantial displacement of the land masses, which had an influence to shift the position of the Earth rotation axes up to perhaps 10 degrees towards to the area of Spitsbergen. Even the tsunami event in the Indian Ocean at Christmas 2004 led to a slight shift of the rotation axis.

Since the volcanic island of Iceland is located on a hot spot or on a crossing point of three tectonic plates, here methane could have come into contact with magma, whereby the huge volumes of gas could not be oxidized and the original much larger Iceland was literally blown into the air, as a sort of giant lava volcano. In fact, old maps show a much larger Iceland with locations, towns and landscapes, which today can be found destroyed deep down on the ocean ground.

Did several such events take place or can be discerned a connection with the separation of the North American and Eurasian plate, whereby Greenland also supposedly distanced itself from the European continent? This scenario is thought to have supposedly occurred about 55 million years ago, according to the researchers, when at this time a lot of methane and carbon dioxide was discharged and the Earth warmed dramatically in the Arctic (Storey, 2007).

In the newspaper *Hamburger Echo* of 15 September 1951, there are reports on apparently curious findings: "The expedition ship 'Meta' on its last trip to the island of Helgoland was able to make a discovery of inestimable worth. Two dolmens were discovered in a mud bank at a depth of 30 meters in addition to remains of houses, grave accessories, ancient handcraft equipment and other objects of utilizes from the young Stone Age and the Bronze Age"(cited in Meier, 1999, p. 490).

The stormy North Sea is a very young, shallow basin. Geologists assume that this area was filled with debris from Scotland and Scandinavia at an early stage of the "Ice Age", it thus became dry land. Fact is, the Rhine flowed through this dry lowlands, and the mouth of this river was located near Aberdeen in Scotland (see Overeem et al., 2001). The Thames was a tributary river of the Rhine at this time.

In the Bronze Age our ancestors lived in the area of today's North Sea. Throughout Northern Europe and also the at this time dry North Sea were ideal areas of settlements. In any case, the sea has not extended gradually all time, but engulfed the region sometimes, partly with super-massive flooding in the search for new horizons. The Dogger Bank may have towered out of the water for some time, but was finally overrun by the sea.

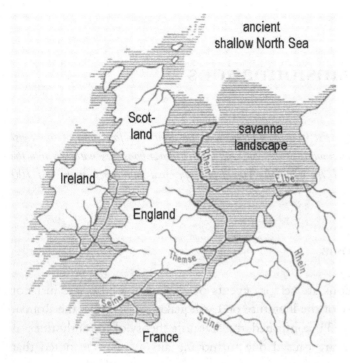

ancient shallow North Sea

savanna landscape

Scot-land

Ireland

England

France

Fig. 17: Dry lowlands.
The shelf areas of the North Sea and around the British Islands in the megalith era and long time after were populated land areas (hatched areas) – at the time as Iceland had not yet subsided. After Bastian, 1959, p. 27.

At a conference in Glasgow (Scotland), the geography professor David Smith of the *University of Coventry* presented his theory; based on 25 centimeters of thick deposits. Great Britain was separated from the European Continent after the "Ice Age" as a result of giant waves to form an maiden-like island (www.wissenschaft.de, 14 September 2001). With this flooding, and the separation of Great Britain and Ireland from continental Europe a drastic climate disturbance resulted that was accompanied by a violent tectonic faulting, as described in my German book "Kolumbus kam als Letzter" (Zillmer, 2004, p. 282 ff, see Hsu, 2000, p. 174).

2 Landscape Transformations

About 1100 mud volcanoes are active contemporaneously on the Earth so far, some of this exist submarine. These spit out daily huge amounts of methane and other hydrocarbons very naturally into the atmosphere. Off the coast of Trinidad, the Chatham Island has appeared four times in the last 100 years, as a result of mud volcanic activity.

Seismic Remote Diagnosis

Is it possible that gas eruptions are seldom events or quite simple, they are just not recognized as such? In a review of the literature on earthquakes of his time, the Roman writer Annaeus Seneca (−54 till +39) confirmed, the favourite theory of the authorities of that time was that earthquakes are caused due to moving air (= gas). He noted that "Before the earthquake it is usual to hear a roaring noise, from winds that created a subterranean disturbance" and that often when an earthquake occurs, and only a part of the Earth's surface is broken, wind blows from there for several days, as it was the case during the Chalkis (Chalcis) earthquake that struck eastern Greece . . .

This is not an individual opinion, but the descriptions of the eyewitnesses at that time from many parts of the world, where was reported of similar observations in relation to earthquakes: thunderous and sizzling sounds, sulphurous fumes, changes in ground water, hot gases and flames. The British seismologist Robert Mallet published a 600-page catalogue with descriptions of more than 5000 historical earthquakes. It describes accompanying phenomena, including light phenomena (earthquake lights), explosions and special atmospheric phenomena. John Milne (1850–1913), the inventor of the modern Seismographs, came in his textbook to the conclusion that most earthquakes were caused through the explosive action of steam in a process that was connected with volcanism.

During the 1906 earthquake, which destroyed some parts of San Francisco and all of Santa Rosa, it was observed that the shocks were preceded by a roar and significant wind. In the Santa Rosa Democrat-Republican on 23 April, it was reported that there were gaping cracks caused by the earthquake in the Earth's crust from which strong gases were being emitted that made people and livestock ill. Later, on 4th April 1910, the newspaper reported that two nights before the earthquake there were small lightning flashes that streaked on the ground. These were reported 30 hours before the earthquake in an area which was later to become the epic centre. It is therefore not surprising that there is a reported about abnormal behaviour of animals that happened before earthquakes and this indeed from all parts of the world – in ancient as well as in modern times. Alexander Humboldt described in 1822, that people who fear earthquakes tended to attentively watch the behavior of dogs, goats or pigs. He could not decide whether the animals can hear underground sounds or smell the exit of gases, whereby he made it clear that the latter possibility cannot be denied.

Animals feel precursor phenomena of earthquakes, but why earthquake experts are able to make predictions only in exceptional cases (by chance?)? In contrast to earlier geologists, modern geologists and geophysicists work at a computer and only occasionally participate in a field trip, if research funds are available. They therefore sit far away from its epic centre, and watch the eruption of tension and tilt measurements. This scientists count on things, which (supposedly) are scientifically understood. Precursor phenomena are scientifically unacknowledged nowadays, because one believes that earthquakes can simply be explained through (tectonic) tensions.

Since geophysical statements are based solely on seismic measurements, it is assumed that at a certain place through "elastic storage" of tensions the breaking point is exceeded and as a result of the breakage the excessive stress is then transferred to another location. Thus a greater area around should fractures in quick succession, almost as a kind of domino effect. In fact, the process is more complicated from a soil mechanics perspective, since in the absence of a pore fluid, the rock take on plastic characteristics and will not suddenly break in the form of a brittle fracture, once a critical state of stress is exceeded. A procedure that has been thoroughly researched through fracture experiments with concrete structures and serves as the basis of static calculation methods. Since geophysicists only compile colourful sketches into patient computers, but do no study science of strengths of materials or even take not into account effective material properties, there is no pure mathematical proof or even a laboratory testing on supposedly "stored elastic tensions". Rocks similar to concrete layers can absorb compressive stresses, but on the other hand they can hardly absorb tensile stresses (tension). An "elastic storage" of tension is not possible in this way, since the rock would tear due to the hindered friction and thus the storage potential would be lost.

With suddenly overstepping of a breaking point many of the recently documented phenomena cannot be explained by the geophysicists. In effect, shortly before the Earth shakes, less low-frequency subterranean radio waves are emitted than it is normally the case. The intensity significantly decreases up to four hours before a night time earthquake. This was discovered by French researchers in 2008 through the evaluation of more than 9000 strong earthquakes, which had originated from less than 40 kilometres below the Earth's surface. The values measured by satellites are used to compile a map of electromagnetic radiation. On this, the (fictional) "elastically" plate tectonically shift processes stored before the fracture can exert no influence at all. But animals can nevertheless still perceive these precursor phenomena in the form of changes in the radio-wave emissions that serves to explain the often observed almost prophetic anticipation of an earthquake (Němec et al., 2008).

Chinese researchers from the *Nanyang Normal University* in Henan, discovered unusual cloud gaps several weeks in advance of two major earthquakes in the southern Iran, although the surrounding clouds were moving. At the same time, the scientists observed in both cases, an increase in soil temperature along the fault lines. Already in the 1980s, Russian scientists had observed changes in temperature and unusual cloud formations before earthquakes occurred. Although this phenomena arose along fault lines, this has nothing to do with plate tectonics, and there is also no attempt to come up with conventional patterns of explanation.

That's why earthquake researcher Mike Blanpied of the *U.S. Geological Survey* is mindful: There is no physical model with which one could explain why two months before an earthquake something unusual suddenly occurs and immediately again disappears, without returning again. The Chinese researchers and myself are however emphatically of the opinion that the clouds were dispelled by the gases, that escaped out of the fault lines. They further assert that this would also provide an explanation for the increase of the temperature in this area (Guo / Wang, 2008).

On the one hand gases can slowly escape over a longer time period or on the other hand explosively blow out (Gold, 1999, p. 144 f.):

"After the puff of fluid into the atmosphere, the pore spaces that had been created in transit may collapse; such a collapse offers a sound explanation for the vertical displacement of chunks of crust during earthquakes and for the volumetric changes in sea floor or continental shelf that would be needed to induce tsunamis. In the great Alaskan earthquake of March 28, 1964, for example, some stretches of land sank within seconds by as much as 30 feet. Presumably this means that the ground below suddenly became denser. But rocks are not compressible to such extend, nor would such compressions occur suddenly. Pore spaces that had expanded the rock with high-pressure gas must have been involved, and when the gas abruptly found an escape route, the pore collapsed. No fluid other than a gas could have supported the rock and then got out of the way in seconds. Similar events have been recorded in many historical earthquakes", Thomas Gold describes his theory as *gas emission and elastic rebound.*

In other words, as a result of a gas explosion occurs an earthquake, and it opens a crack in the Earth crust, which is closed again after the explosion, leaving no marks. Maybe, nothing is to see and nobody comes to the idea that there was an opening in the ground before. In contrast, Western scientists believe "that earthquakes are of purely tectonic origin, caused by an increase of stresses in the rock" (ibid, p. 145), as a result of tectonic (mini-)shifts; if the tensions overstep the breaking point of the rock debris.

In case of a gas eruption it behaves differently. The generation of many small cracks *reduces* the breaking point of the rock, until a critical value is reached. This scenario is well known and researched with respect to concrete constructions. Nevertheless the geophysical viewpoint is that an "elastic" storage and an increase in the "tectonic tensions" occurs, until the breaking point of the rock is exceeded. These are two very contrasting explanatory models! These lead to very different worldviews about the structure and composition of the Earth's crust or lithosphere, indeed of our entire planet.

If rapidly rising gas is the cause of an earthquake, there can be no prediction of earthquakes by reading of seismic instruments, since it is really only the pore space of the rock that is filled with gas and there may be only a bulge could be measured at the surface. As a result, there is also the nowadays scientifically neglected precursor phenomena, which the animals reveal in case of their behaviour, for example, by smelling gas leaking form the Earth's crust or by perceiving sounds generated at frequencies that are not audible for us. These animals would not notice the build-up of tectonic tensions in the ground respectively rock debris.

The theory of rising gas can be used also to explain eloquently other precursor phenomena, among others the in the meantime documented turbidity of groundwater before the earthquake occurs. Similarly, multiple quakes or aftershocks are plausible, because with the rise of the same amount of gas can be triggered several earthquakes on the way up, because several superimposed layers are broken one by one (Gold, 1999, p. 141 ff.).

In connection with volcanoes it is possible to explained, why some volcanoes keep quite often over long time periods, and then to become active suddenly. Could gas be the "ignition", when it penetrates into lava vents? Lava, but also mud volcanoes become suddenly active, in the aftermath of precursor and earthquake phenomena. Actually one measures gases by eruptions of lava volcanoes, however with differing gas compositions. In general, water vapour and carbon dioxide are the dominant gases; however, often a relative percentage amount of methane and hydrogen is registered. That seems too little for the previously outlined degassing theory.

However, there are two different degassing scenarios to see. Gases in a slow transport process through existing magma in solution or in the form of small bubbles form a chemical equilibrium with the magma. These have to deliver oxygen for the oxidation of hydrogen or methane. By this way the oxidation of hydrocarbon in turn leads to the creating of water and carbon dioxide. Consequently, the chemical composition of the magma regulates the amount of the remaining non oxidized gases, in consequence of the oxygen content. It also becomes clear why water often escapes during volcanic eruptions and partly water clouds infiltrated with dust are blown into the atmosphere.

So it was a hardly noticed sensation by the volcanic eruptions of the Etna on August 1 and 2, 2001, as these spit water clouds – a phenomenon only perceived by experts.

*Fig. 18: **Water vapour. Picture left:** White water vapour clouds appereared during the Mount Etna eruption in Sicily in 2006. Is the thick dark smoke related to the burning of higher hydrocarbons? See photo 27. **Picture right:** The emergence of the Paricutin ash cone and escaping dark and white smoke in Nicaragua in 1943.*

For this scenario there is no explanation if we do not take into account the hydrocarbons existing in the substrata respectively migrating upward to the surface.

The water released during volcanic eruptions that did not exist before, was chemically originated inside the Earth and transported to the Earth's surface through the volcanic eruptions in the form of water vapour and mudflows. For geophysics it is however a mystery, why so much water is released during volcanic eruptions, because the present viewpoint is that there occur no new formation of water, but rather a closed cycle, since the model of the hot Earth does not allow a geochemical creating of virgin water.

Therefore, it is also officially not discussed, where the 1,000 cubic kilometres water came from which were emitted from the volcano Mount Tambora in 1815. The still active volcano Shiveluch (Sheveluch) on the Russia's Kamchatka Peninsula has hitherto supposedly spewed up to 4500 cubic kilometres of water in the atmosphere (Drujanow, 1984, p. 59). In the vicinity there are several volcanoes, such as the Klyuchevskaya Sopka – Eurasia's largest volcano – and in a valley about 90 geysers exist, from which water fountains shoot up to a height of 40 meters.

With such intense bursts of water out of volcanoes a greater part of the hydrocarbons is also oxidized, whereby also carbon dioxide is originated. If, however, methane and hydrogen may not have been fully oxidized by the oxygen present in the magma, the amount of gas which races through the magma, can be so large that the generated gas mixture dominates and the volcanoes explodes. That magma itself is not explosive and therefore was not the cause of eruptions.

By a major explosive volcano eruption the gases blown in to the atmosphere would have to leave traces. In fact, there are often eyewitness reports of flames or even lightning timing during the time period of the volcano eruption, however also similar observations shortly before or after the eruption. In addition to flaming lava that are a common phenomenon for example, on Iceland or Hawaii, sometimes high flames shoot out of a volcano vent during an eruption.

There are descriptions of flames from the Mount Tambora eruption of 1815 (Raffles, 1817, p. 28), the eruptions of Santorini 1866 as well as Mount Pelée in 1902 and an eruption of the Fuego Volcano in Guatemala, 1974 (Dawson, 1981). A more detailed description of flames exists also from the Krakatoa explosion in the Sunda Strait between the islands of Java and Sumatra in Indonesia in 1928. After several days underwater volcanic eruptions, orange-yellow flames appeared, dancing on the surface of the water above the crater. Observed "from a distance of about 200 meters, that flames were about 10 metres high" (Stehn, 1929). Confusion with glowing ash is excluded in this case.

Overall, there is no typical mixture of volcanoes, because gas is rising upward from the mantle in quite different composition and become oxidized– at volcanoes , with different intensity depending on the speed of the gas flow through the magma. Let us see now what happens if gas makes its way through cracks and fracture formations into a magma chamber. When the amount of oxygen provided by the magma is not sufficient for the oxidation of the gases, these reach the channels leading upwards. As the pressure around the magma with less depth strongly decreases towards the top, the non-oxidized gas tends to race to the top. Thereby accelerating the growth of the gas volume, and the expansion is very severe near the surface. The overlying lava can then not form a kind of plug like cap, but rather gas fingers pierce through it and produce flames when ignited, while the resulting turbulent mixture of gas and lava explosively shoots out of the volcanic vent and enormous quantities of ash are ejected.

Eruptive gases are the main acting force of volcanic activity! This described scenario is also known in case of deep drillings. The sudden increase in drilling fluid can be a danger signal, and the hole must be closed, since otherwise too much mud would be ejected and a violent explosion would follow. In this case, a gas bubble races from the bottom to the top, enormously expands and ejects more and more mud. The expansion becomes explosive when the gas bubble suddenly expands in the atmosphere after the drilling fluid was ejected. Something similar happens at volcanoes, and that is why one finds a large caldera (*Spanish* boilers) after intense eruptions, caused by the collapse of the surrounding rock. But there is however a difference! On the one side in case of deep drillings there is coming mud or sand out of the borehole, and however, on the other side in case of lava volcanoes just lava occurs instead of sand or mud. But for both forms of appearances, however, was found the same reason: rising gas, normally methane or in the event of oxidation just carbon dioxide (and water or vapour). Consequently, mud volcanoes must also exist analogous to lava volcanoes, and indeed there is such an eruption phenomenon that is little known but very interesting.

Mysterious Mud Volcanoes

On 29 May 2006 in the eastern Indonesian island of Java started a mud volcano to erupt.. Huge quantities of petroleum-containing mud were ejected in 2006: 50,000 cubic meters in June, 126 000 in September and 176 000 in December. Yet in May 2007 there were 100,000 cubic meters, an amount sufficient to fill 50 Olympic swimming pools. It all began with a mud-explosion near an oil borehole in which the mud was shot 50 meters high into the air, while methane and hydrogen sulphides were released from a crater 60 meters in diameter.

The conventional explanation is always the same. Because earthquakes and volcanic eruptions are generally consecutive symptoms of tectonic events, the activator for this mud volcano in Java should have been an earthquake with a magnitude of 6.6 on the Richter scale, the second earthquake event before the mud volcano break-out. This shook the area around Yogyakarta, where 6,000 people were killed. Was an earthquake the actual reason or it is possible that the mud eruption was caused by upward rising gas? In such a way a different theory declare that there happened a mistake during the oil drilling on 28th May, one day before the "explosion". At that time occurred an uncontrolled break-in of an unknown liquid into the borehole. The steadily – with high-pressure pumps – in the chisel pressed and at the end of the borehole squeezed out water–clay mixture suddenly went lost. Normally this drilling fluid takes along to the Earth's surface the dissolved rock particles. A continuation of the drilling was no longer possible and the borehole was sealed at a depth of 643 meters. Witnesses confirmed that the mud didn't come up the next day not from the hole, but 150 to 500 meters away (*Jakarta Post Online*, 16 June 2006).

Mud volcanoes exist in many oil areas, like at the north shore of the Black Sea, in the southwest of the Caspian Sea, in northern Iran, in Romania, in Burma at Minbu, on the Indonesian island arc, along the Gulf Coast of Texas and Louisiana and other countries. The largest mud cones are near Baku on the Caspian Sea and in Burma. The 300 meter high ejected mud in Baku is mixed with asphalt (Stutzer, 1931, p. 281).

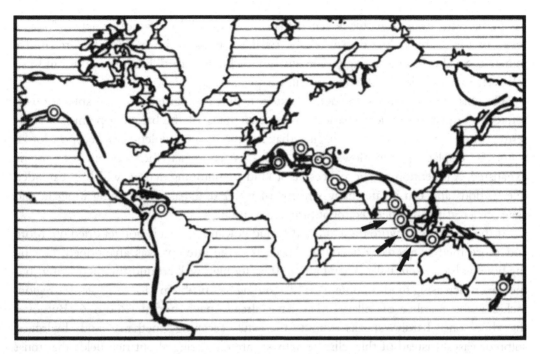

Fig. 19: Fold Belt. *"The black lines show geologically young faults and fold belts. The 13 most prominent mud volcano areas are marked and they all fall on these belts"(Gold, 1987, p. 101). In these areas such as along the Indonesian island arc respectively over the course of the volcano and earthquake belt (parallel lined to the alleged subduction zone and deep sea trench – see black arrows) there is an abundantly production of crude oil and natural gas to register.*

On the coast of Burma in1907, an Island was formed as a result of a solid discharge of mud: 400 meters long, 150 meters wide and 7 meters high. On the south coast of Trinidad occurred a submarine eruption in 1911. Rock material and gases were blown out of the lake. The gas ignited, and a 30-meter-high flames rose. At the same time an island rose from the sea, five feet above sea level.

Tectonic shifts are not the actuator for active mud volcanoes, although the thirteen best known mud volcano areas lie on geologically young (Mesozoic and Tertiary) faults and fold belts. The secret is to see in case of upward migrating gases which can rise easier through cracks and fracture zones in the Earth's crust (Figure 19).

The mud eruption in Indonesia was apparently triggered in consequence of drilling a borehole, where gas was responsible for the discharge of crude oil-containing mud. This reached a temperature of about 100 degrees Celsius in Java. Although it is definitely be fixed that methane pass out this mud volcano to this day, it is only taken as a kind of inexplicable side note. The geological argument doesn't take into account the involved gas in any way, instead of this is stated that the so-called "cold volcanic activity" was a result of the relatively low density in the Earth's crust and the swelling feature of clay minerals. So the 100-meter high fountain cannot be explained! High pressures and not relatively low densities are the only possible explanation.

Why this mud volcano is to such an extent active, geologists have no explanation by reference to plate tectonic scenarios. The overstepping of the breaking point is a short term event. However, one can imagine that for the case of violent gas injection lot of water is displaced, but flows back again as soon as the amount of gas is exhausted. A repetition of the process is a kind of pumping action, which is extended over periods and can create many streams of water until gas is streaming in again and again, whereat fine-grained sediment turned into mud.

At the time when I write these lines, the mud volcano in Java is not yet dried up. More and more mud is ejected. Already in April 2007 the attempt was abandoned, to clog the hole with concrete balls. They built a 40 meter high wall around the mud crater, but the ever-increasing mud-mountain is becoming a problem for the local region, including the number of villages that have already been buried (Photo 6). But over the while one hardly speaks about the still escaping methane and hydrogen sulphide gases. Instead of this, climate activists are discussing about rice fields that must to be closed, because (very) little amounts of methane are leaking. To stop this it shall be taken away the food foundation of this populations in Asia – but no one is measuring the amounts of leaking (abiogenic) gas of mud volcanoes and other sources.

The principle of this mud volcanoes recall the newly discovered Martian geysers with several kilometres high fountains of muddy water, springing up from the Martian surface and causing deep valleys, landslides and discharges on the Martian surface. Mud volcanoes can arise anywhere where gas and water can be pushed upward through soft rock (clay) or sand: The result is upcoming mud and the water is usually salty. If the pressure of the accumulated amounts of gas increases in the underground then there can occur a sudden eruption and a mud volcano is born. It is sometimes a lot of mud ejected, and the gas can ignite.

This principle also works not only at the edges of tectonic plates. In the case of gas and deep oil drillings it can be happen that instead of oil or gas big masses of mud or sand are ejected similar like a regular mud volcano. For example, the output of some boreholes at the Sunset Field in California consist to two-thirds of sand. A single borehole lifted 20 000 cubic meters of sand in two years.

Some pump wells in the North Midway Field in California produced about 35 000 cubic meters of sand per year. Even in Russia one knows a spouter (gushing spring or springer), in which oil (or sometimes including water) jumps out of the borehole. There, in addition to hundreds of thousands of cubic meters of gas and oil, a lot of sand and stones are thrown out. At Bibi Eibat a spouter spat 10,000 tonnes of oil and 10,000 tonnes of sand in only one day (Stutzer, 1931, p. 283).

The driving force of an oil spouter is the pressure of the gas. This cause must be underlined. We know definitely that methane and other hydrocarbons fumed into the atmosphere out of mud volcano craters, but this circumstance is ignored to explain the activity of "cold volcanoes" and is keep secret in the climate debate. By taking into account the rising gases from the deep we find the same cause and thus unified explanation for both types of volcanoes – whether "hot" or "cold" volcanoes. The difference is only in the rate of oxidation of rising hydrocarbons. Therefore especially mud volcanoes (= cold volcanoes) are emitting large amounts of gas. If not already done so far, in the vicinity of mud volcanoes one should drill for gas. The question is not whether you get methane but instead of this only whether the quantity is sufficient for economically productive use.

Mud Volcanoes of course, cannot run as deep and be as explosive as the lava volcanoes, since the pressure in the mud doesn't change so rapidly in the depth as it does in case of magma. In addition, the gases in mud channels are much cooler than in lava. Nevertheless, the two types of eruption show many similarities because both types of volcanoes probably caused by the same process, namely, the sudden rise of gas that is deeply injected and has been pushed up through long, fluid-filled channels. Therefore "allow a gas bubble to ascend a long way, and to expand greatly, producing the explore eruption" (Gold, 1987, p. 102).

Lava volcanoes emit mainly water vapour and carbon dioxide and a little less methane, if the oxygen supply of lava is consumed with the oxidation. Mud volcanoes however, emit mainly methane, but also other hydrocarbons, because the oxidation of these gases as this occur in case of lava volcanoes is omitted. If one considers that the vaporized water in the depth is the engine of mud volcanoes, then it is not taken into account that the steam would condense on the way up to the Earth's surface. Hot water is relevant *only at shallow* mud pots, since the system heats the water in geothermal active areas.

It is therefore not surprising that some regions of mud volcano are located near lava volcanoes, like in Alaska. The carbon containing gases ascending from the depths become oxidized on the one hand, if these come in contact with lava in *small* amounts and on the other hand remain basically unchanged, if the upward rising takes place through the cooler channels of mud volcanoes. Therefore we can state that there is a common principle of causation, but on the Earth's surface are to see two seemingly totally different phenomena, although hot and cold volcanoes sometimes existing side by side in some areas.

Already A. Daubrée experimentally demonstrated that high gas pressure can open channels in the crust on its on (*Nature*, July 6, 1893, vol. 48, P. 226 ff). Now it becomes understandable why mud volcanoes can ascend fire bursts. On 5 January 1887 a 600-meter-high flame was reported from the mud volcano in Lok Botan in Baku area on the

Fig. 20: Mud volcano. Output hole in the hot Norris Geyser Basin on the north western edge of the caldera of Yellowstone volcano in Wyoming.

coast of the Caspian Sea (Stutzer, 1931, p. 281). It poured a mudflow which was 300 meters long, 200 meters wide and on average two meters thick.

Almost 100 years later, a flame shot up even to 2000 meters in height, and then with lesser height still burning on eight hours.. The diameter of the crater was 120 meters (Sokolov et al., 1968). It is estimated that for the eruption of gas is required approximately one million tonnes. Indeed a huge amount! In this area these types of eruptions happen every decade. According to calculations of measured data about the gas to mud ratio, solely for this big flame event it has burnt more gas than it is contained in one of the largest commercially exploited gas fields. The occurring gas amounts in the depths of the Earth have to be tremendous!

Worldwide, there are now 1100 active mud volcanoes of which some are submarine active. These spit huge amounts of methane and other hydrocarbons on a daily basis naturally into the atmosphere. The island Chatham before Trinidad appeared as a result of mud volcanic activity four times in the last 100 years. The diameter of such emerging islands can be several kilometres or even some meters.

In the depths of 1270 meters in the Arctic deep–sea off the Norwegian coast is situated *Håkon Mosby* – a mud volcano with a diameter of about 950 meters – on the continental slope in the western part of the Barents Sea. The mud cone has risen about 12 meters above the seabed. The outgoing mud about 28 degrees warmer than the sea water at that depth. Here rise large quantities of gas, consisting of 99 percent methane. This gas escapes into the atmosphere, of which a part swirled in sea water by fast currents and is oxidized by oxygen to originate water and carbon dioxide in the ocean.

Another part of the escaping methane is processed by bacteria living on the seabed. This process is called "anaerobic oxidation" and shall work up of only 40 percent of the rising methane on *Håkon Mosby*. Until now, scientists assumed that in areas with a high flow rate of methane are living also significantly more methane-eating micro-life organisms live, was confirmed in a press release from the *Max Planck Institute for Marine Microbiology* on 18th October 2006. In other words, it is believed that there is no methane entering the atmosphere, because the huge amount of methane shall be spread by deep currents over large distances in the sea water and partly quickly converts into carbon dioxide by bacteria. This remains in solution in the ocean water or can form carbonates.

Methane itself is neutrally charged and not readily soluble in water, but oxidized, whereby carbon dioxide and water originates. Such an event increases sea levels and the carbon dioxide content of the water. The carbon dioxide is dissolved in water and at an increased warming of the oceans emitted into the atmosphere. Therefore, at higher temperatures releases the oceans naturally huge amounts of carbon dioxide, and saves in turn again at lower temperatures. In addition, from about 1100 known active mud volcanoes large amounts of greenhouse gases are naturally released throughout into the atmosphere. But there is still a different kind of "volcanic activity", which liberates directly methane and other hydrocarbons.

Not only lava volcanoes (mainly carbon dioxide) and mud volcanoes (mainly methane) releases large amounts of "greenhouse gases" on a daily basis, but there are vents in the crust, from which no lava or mud leaks, but in addition to other hydrocarbons mainly methane is released directly. It is so far not discussed, not even really perceived as a third "volcano-type", which can occur exactly where neither magma nor fine-grained sediment is present in large quantities in the Earth's crust.

Such a methane volcano artificially can be brought to outbreak analogous to the mud volcano, for example like in Java by drilling in a bubble with over-pressured gas. Such an event happened – as has been already described – as an oil rig in the North Sea was destroyed in 1990. Powered by the methane in the bubble hole, the pressure burst out through the borehole and was vented in a violent explosion. The oil rig was destroyed like the Louisiana Gulf Coast Deepwater Horizon oil rig in 2010.

From Pockmarks and Pingos

The Earth seems to be degassing and let methane expand through the Earth's crust from deep geological layers everywhere, where cracks, fractures or weak zones are present and not enough oxygen for its oxidation is available or where no methane hydrate can be formed. It has been noted that the East Pacific Rise, a mid-oceanic ridge, has methane directly emitted along with very hot water over a large part of its length (Kim et al., 1983). Because it is too hot along this crust-crack no methane hydrate can be formed. Also in the rift of the Red Sea is coming up methane along with hot water in the depth, in the vicinity of where oil exists in commercially usable quantities.

In shallow seas there is absent the necessary pressure to form methane hydrate. There, the methane can rise directly through cracks and holes in the ocean crust into the sea water and even into the atmosphere. If fine-grained sediments existing in the substratum then it is possible to build methane-emitting pockmarks and mud volcanoes, which are

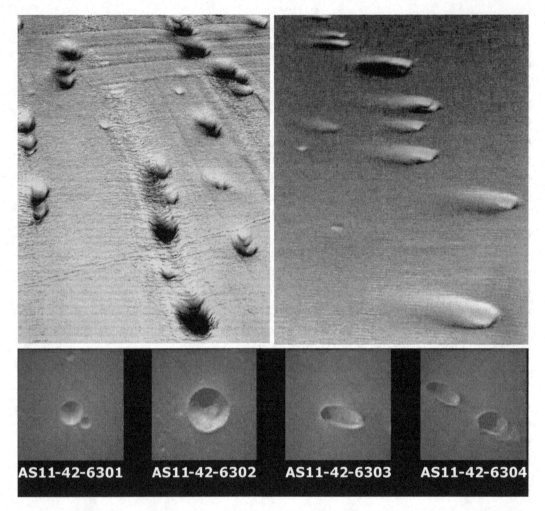

AS11-42-6301   AS11-42-6302   AS11-42-6303   AS11-42-6304

*Fig. 21: **Pockmark fields. Upper pictures**: Typical pockmark with a 10 to 15 m diameter at the bottom of the North Sea: Picture upper left symmetrical and upper-right asymmetric formations. The orientation in some areas is similar, but also varies from area to area (Judd / Hovland, 2007, p. 12 and 14). **Lower figures**: Do these (allegedly) Apollo-11 pictures show pock-marks on the Lunar surface which arose from gas leaks?*

found naturally ashore. In areas where the Earth's crust has impermeable layers as thick sealing caps the rising methane under this layer is stored with increasing volumes, as in the permafrost of Siberia.

Therefore, methane must degas everywhere where the Earth's crust is porous or has fracture zones. In such areas as already discussed pockmarks appear on the seabed. This sea-water above them is considerably enriched with methane. With the help of sonar it was discerned that pockmarks fields have a layered substratum structure, which point to a continuous degassing process; because with smaller methane gas explosions new fine sediments are continuously thrown from the substratum into the sea water, which then accumulate in layers on the seabed on upon the other (Figure 15, p. 43).

If the pressure is not sufficient to allow the gas to burst-out explosively and to leave behind crater-like pockmarks, then swell mounds as a result of the bubble building in the substratum. These caverns can be filled with gas, ice or fine-grained sediments. Such structures are to find in permafrost soils of higher latitudes, also in North America and northern Siberia, on Greenland and Alaska, in Spitsbergen as well as in the Antarctica. There mounds arise isolated from each other, sometimes perfectly grouped into fields, which have diameters from 6 up to 1000 meters with a height of up to 50 meters. This largely ignored phenomenon is called pingo that in the language of the Inuit simply refers to a mound.

In the core of pingos are often exist ice. How do evolve such up to 50 meters high ice cores so that the ground bulges to form a mound. Although the inflow of water is limited in permafrost areas, the ice should be formed from the liquid and gaseous water existent in the *immediate* environment and the substratum. It is thus hardly explainable why the pingo has an average vertical growth of up to 20 centimeters per year that requires a continuous process. Can large pingos constantly feed themselves even with water from the surrounding permafrost (Figure 22) ?

It is yet officially discussed another pingo type that shall be powered by a hydrostatic system with supposedly unlimited water supply as a result of groundwater flowing from an outside source – called open-system. One imagine a water source *hat do not freeze in the winter time and do not run empty in summer time*. This should ensure a permanent water supply to be ensured – in permafrost areas! In this way one could imagine a steadily growth of the ice, although a year-round supply of water in the permafrost area would hardly seem possible or only in exceptional case.

Nor is it plausible to divide a very common, even typical landscape form in the permafrost into two or even three special cases with very different attempts to explain (Wiegand, 1965, p. 9). Is there perhaps a sole cause? Can this be found in the accumulating gas underneath the permafrost that migrates upwards through fissures and cracks? Near the surface ice plug can then evolve as a result of the Joule-Thomson effect in case of relaxation and an increase in volume of the gas near the surface of the Earth to come along with a drastically cooling of the surroundings – like the atmosphere and the ground and as may be the case with a water stock which shall be freeze also (see Figure 10). The water can be formed as a result of the oxidation of methane or be pumped upwards through the aforementioned mud volcano principle. This pumping process is repeated in periodic intervals, which explains the puzzling seemingly continuous growth of a pingo. The filling of the frosted mounds occur thus similar to the volcano principle,

*Fig. 22: **Pingo.** **Left:** Crater lakes and pingos near Tuktoyaktuk in Northwestern Canada (Photo: Emma Pike). **Right:** Exposed-pingo ice body.*

dependent on what kind of material is present in the substratum. Thus, if there are deep fine-grained sediments present near the surface, these sediments are injected as water-soil mixtures in the pingo-cavity, otherwise the pure water is frozen through a drastic cooling as a result of the Joule Thomson effect. If there is no ice formed, the locals use such methane sources near the surface as natural cooker or by larger gas quantities a pingo mound evolve with a gas filled cavity.

In the event of overpressure the degassing can take place explosively in the form of small methane volcanoes like the pockmarks on the seabed. Pingos are therefore also called "ice-volcanoes". Huge ice-volcanoes exist among others on Saturn's moon Enceladus. The Cassini spacecraft in March 2008 flew through a 50 kilometers high fountain of gas and ice, which spouted out of the only 500 kilometers large ice moon. The composition of the gas fountain will remain a secret, since at this moment, oddly enough, the measurement instrument stopped working. Had one measured methane? To the surprise of the researchers, from a series of columns on the south pole of the ice moon, allegedly ice, water vapor and dust was sprayed out with a speed of 400 meters per second. The interior of Enceladus must be much warmer as the outer space (Photo 31).

Several authors have made reports of explosive eruptions of recent pingos also on Earth (Strugov, 1955, p. 117), also from ice-volcanoes. By such explosions up to three cubic meters of large blocks of rock and ice have been catapulted over a wide area. "The destruction took place exclusively at the peak of the mounds" (Wiegand, 1965, P. 41).

This description, however only involves small frozen mounds (pingos). But if – as in the *Teufelskaute* in Germany – larger quantities of gas are trapped and pressed under enormously pressure inside larger pingos with a diameter of 60 to 100 meter. This incredible pressure could lead to a violent explosion and to build a crater (ibid., p. 43).

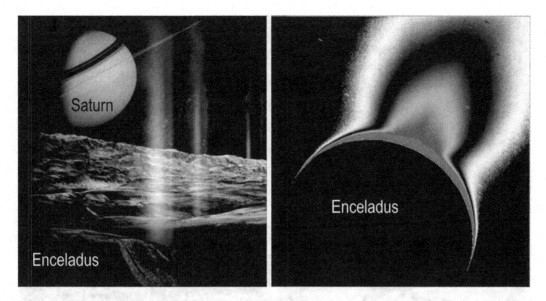

Fig. 23: Gas jets. Left: *At the south pole of Saturn's moons Enceladusdas there are fountains of "water ice" that are active.* ***Right:*** *The color-edited recording of the whole spectrum of the outbreak. In the course of a fly–over the Cassini spacecraft analyzed the water fountain; it was twice as dense as the gas that surrounded each individual jet. NASA / JPL / Space Science Institute.*

The aforesaid *Teufelskaute* – that also shall be deemed to be a ritual site – is today a small crater lake in the Fulda region in Hesse, Germany, marked by a prominent wall on the east and west side. This is probably a fossil pingo, as Gottfried von Wiegand (1965) described in detail. There are many other such structures in Central Europe, which are often not recognized as fossil pingos. Through excavations it has been shown that in fossil pingos existent rocks are bolted upright at 60 degrees toward the center of the pothole – like a caldera of volcano, but even not like an impact crater .

Pingo are naturally a temporary phenomenon. Their subsequent landforms – like buried, transformed or water filled hollows or potholes – are only seldom recognized. Not only in Siberia, but also in Central Europe once existed a permafrost formation, and therefore pingos are also often found in Germany. These have been wrongly interpreted as roman camps, bomb-craters, water reservoirs, clay pits, simply as surface depressions or old pit houses, a dwelling dug into the ground (Paret, 1946, p. 54-83).

A parallel relationship between pingos and methane or mud volcanoes is now obvious. The gas rising from below through the cracks and fracture zones accumulates below the permafrost. This is cambered by the gas pressure to build a mound. Such pressurized "air spaces" are often found in the interior of some bulging hills in Siberia, confirmed already Wiegand (1965, p. 41, see Müller, 1959, P. 67).

Fig. 24: Fossil Pingo. Upper left: *The Hasenkaute is located in the vicinity of the Teufelskaute between Hemmen and Großenlüder in Hesse, Germany. This is not an impact crater, but a fossil pingo (Wiegand, 1965, p. 17 ff).* ***Upper right:*** *The erect bedrock at a 60 degree angle toward the center of the Hasenkaute due to the pingo formation (ibid., p. 147) – opposite to the expected downward direction in an impact crater.* ***Bottom left:*** *The crater lake-pingo in central East Greenland with a diameter of 100 m (ibid., p. 145).* ***Bottom right:*** *"Witches' rings" on Greenland.*

If enough pressure is built up, ice or gas volcanoes are formed and the permafrost cover is blown away. Subsequently is to see a caldera, mostly described as impact crater.

Can this pattern be used as an explanation for the strange "Witches' rings" of stone, which one for example finds on Greenland? The scientific explanation is that: "The constant change in volume by alternate thawing and freezing over a long time span forces the coarser material to move towards the surface and pushes these to the exterior in circles"(Chorlton, 1983, p. 33). The structure of the "Witches' rings" could easily be explained by rising gas with soil liquefaction and fountain formation (Figure 24).

Because ice is not resistant, a pingo mound disappears in turn. Often unrecognized, in today's temperate zones like Central Europe, fossil pingos can be found, which require significant investigations in order to identify such pingos: The subsequent thawing and slumping down lead to the building of little round ponds and similar surface formations (e.g. in Holland, Belgium, France or England), which one attributes to fossil pingos (Schwarzbach, 1993, p. 46).

Methane rising from the depths of the Earth seems to take place everywhere, if there are weakness zones or cracks in the Earth's crust. Therefore, a thufur called land form is a small knoll or mound above ground, mostly less than one meter in height above the general level of an ice-field. They tend to appear in groups or fields. The icy core of a thufur is distorted and consisted of a mineral filler (soil). This land form can be attributed to the activity of upstreaming methane. Analogously the drumlin formation, which are regarded as evidence for the existence of the Ice Age, can be explained by the degassing theory.

Misinterpretation Drumlins

These elongated, drop-shaped hills are often fan-shaped and staggered arranged in drumlin fields of similar shaped, sized and orientated mounds, representing a drumlin landscape. One is forced to think that drumlins look like pingos, because they have similar proportions, except that drumlins mostly have a deformed streamline. Fossil drumlins are often difficult to identify, but they are filled with clay, sand, gravel and stones – like plenty of pingos. This is also described as glacial till and to be considered to be ice age moraine material (Photo 22–24).

The Dutch geologist Christopher Sandberg writes: "If one examines the drumlin appearance, one must thereby conclude that their *characteristic* features to contemplate in general and in detail point no reference to a glacial origin: The typical shape and almost uniform height of the hills, the almost lowland character of the related planes, the marked morphological connection between both, the segregation of the coarse elements in the hills and the fine elements in the planes, the systematic arrangement of hills in the direction of the slope – *all of these characteristics describe the drumlin, as well as (. . .) the depositional product of water and gas saturated rock streams, which must have had a fast movement shortly before the embodiment of the shape appearance*" (Sandberg, 1937, p. 14 f.).

The geologist Sandberg confirmed that drumlins are no relics of the Ice Age. Although the glacial origin of these hills is continuously emphasized, one can read in the German *Geologisches Woerterbuch* (Geological Dictionary): "The basic cause of origin is still controversial" (Murawski / Meyer, 1998, p. 43). Existing hammocks or sediments, such as older till layers, were supposedly "run over" by moving glacier masses and thereby has been streamlined to be deformed accordingly to the conventional opinion. Can arise in such entire drumlin landscapes, which appear to be newly formed systematically. Where do come from the swarm like arranged dot-shaped till deposits on the Earth's surface? One assumes that these tills are moraine like deposits and were released by much older glacier advances. Can be explained in this way the relatively uniform arrangement of drumlins in regular fields?

Such plains are also found in areas that were never glaciated. Sandberg refers to the so-called "10,000 Hill Landscape" near Tasikmalaya on the western Indonesian island of Java. These types of landform were already described in 1925 as drumlin landscape, caused by wet flow side (Escher, 1925). However the reference to drumlin forms was rejected, because the hill was not elongated in the direction of the flow, although all properties of a drumlin otherwise existed, in particular the internal composition of the hills, in contrast to that of the surrounding lowland, the parallel changing layers of these hills and the "moraine material". "There can therefore be not the slightest doubt of the drumlin nature of the 10 000 hills occurrence" (Sandberg, 1937, p. 14).

A discrepancy thus only exists due to the circular and streamlined shapes. There is a tangible reason behind this dispute; because if the rounded drumlins are real drumlins they were not be created by glaciers and in this case drumlins cannot count as a certain proof of an "ice age". When Sandberg suggests a "fairly rapid movement close to the formation", we then ask ourselves what exactly could have moved here, a "furious" glacier or perhaps the underground itself. As already described in my book "The Human History Mistake", there were violent Earth's crust movements in the Alpine region for about only 2000 years ago, as explained as well in the textbook "Post-glacial climate change and Earth's crust movements in Central Europe":

"At the time of our ancestors, the sea level of the existing lakes was rising – such as Lake Constance or the Swiss lakes – with the simultaneously formation of beach ridges and shore terraces accompanied by a significant demolition of all stilt houses and other lakeside dwellings. At this time of climatic deterioration, *the Earth's crust movements reached a particular intensity and led to the formation of new lakes*, near Munich, Toelz and Memmingen. The wind-borne sand and loess formation came to an end during this time period, and the dunes on Lake Constance, Upper Rhine and in other areas successively developed a forest character. These from a scientific investigation confirmed *Earth's crust movement's in the alpine region took place at the time of the Celts* in a phase of the so-called subatlantic time period from 850 until 120 before Christ" (Gams / Nordhagen, 1923, p. 304 f.).

During such a period quakes the Earth and oceanic pockmarks, pingos and drumlins may evolve in significant numbers, falsely interpreted as evidence of a long lasting Ice Age. During Earth's crust movements, gas migrates through cracks and fracture zones in the Earth's crust, and thereby sediments flush out into or on the surface layers. These sediments can be spill out in the form of wet till streams and sorted by grain size to build the filling of drumlins or pingos. If the pressure is higher, the sediments will be ejecting in the form of a mud volcano or in smaller size as a pockmark.

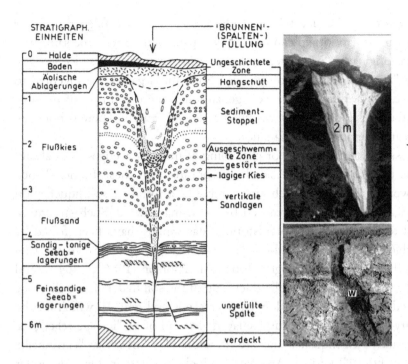

Fig. 25: Earth-quake fountains. *Left picture: cross section of a fossil "earthquake fountain" with soil liquefaction in formerly not glaciated areas in Connecticut. When such fountains are located within the proclaimed inland ice age border, they are then also called "ice wedges" and regarded as evidence of an ice age. After Thorson, 1986.* **Top right picture:** *"Ice wedges" are partially filled with ice, which according to Karl-Heinz Jacob can occur because of the Joule-Thomson effect (see Figure 10).* **Bottom right picture:** *A small ice wedge (W) in a permafrost layer. If there is more pressure build up, than there can also arise an explosive "ice volcano".*

During the evolvement of the described geological formation, the sudden appearance of water or an onset of soil liquefaction plays an important role. In fact, if earthquakes happened sometimes water seems to appear from nowhere in large quantities. This has been demonstrated by the photographs from NASA in January 2001, such as documented by the Jet Propulsion Laboratory. As a consequence of strong ground motions in the West Indies with a magnitude of 7.7 on the Richter scale occurred soil liquefaction in local areas. At earlier dry places, water gushed out of fine sediments suddenly as a result of the earthquakes. Were these earthquakes caused by the gas that was rising together with the water? If for the formation of drumlins arise in case of soil liquefaction associated with slight earthquakes as a result of rising gas, we can now accordingly explain the phenomenon of ice wedges – that are also regarded as proof of an ice age. It hereby involves It is about "more or less vertical, downward deteriorating wedge-shaped, ice-filled columns in soils of permafrost areas" (Murawski / Meyer, 1998, p. 46). Fossil ice wedges are often filled with loess or clay and were therefore earlier known as loess or clay wedges. In accordance with the ice age theory, the sediment fillings are to be *replaced* the earlier ice fillings. Indeed, there exist ice filled funnels or fissures these days. Their primary origin can be simply explained in exactly the same way as pingos and drumlins originated. Since there is no more ice present in fossil ice wedges, experts also speak of "ice wedge pseudomorphosis". Such mind games are invented to suggest the assumption of the ruling of an Ice Age, although there is no proof possible that the funnel-shaped depression (craters) was filled with ice formerly. There is not even an attempt made to search for another explanation.

If we now simply assume that the loess or clay filling of most ice wedges were always present, which means that no ice filling existed ever before, then these funnel shaped depressions are leftover of fountains which arise simply as a result of soil liquefaction caused by earthquakes, which in turn are caused by gas rising upward. Such structures, which are considered to be a result of "prehistoric" catastrophes, have been found by stone pit outcrops which exist in areas with no "glacial origin" (Figure 25). In the U.S. state of Connecticut, the American Geologist Robert M. Thorson (1986, p. 464 f.) and his team identified, the fountain like hydraulic fracturing of the soil material as well as the necessary soil liquefaction in former waterlogged river sands and gravels, which can be identify by sedimentary ripples of the subsurface layers. Instead of an "ice age relict", ice wedges can be the result of earthquakes and are therefore also known as "earthquake fountains", confirmed the well-known Austrian geologist Professor Dr. Alexander Tollmann (1993, p. 148 f.). If such ice wedges actually contain ice, then the Joule Thomson effect is responsible for the sudden ice building during degassing processes and simultaneous earthquakes.

Earthquake Spots and Earth Liquefaction

There are a number of earthquake reports in the great flood traditions of the peoples around the world. What is repeatedly described is how fractures in the Earth crust evolved during an earthquake, from which water gashed up; as can be observed in today's earthquakes. Already, the famous Austrian geologist Eduard Suess (1885, P. 43 f.) described how the ground water hereby shoots up several metres high from countless scattered points. It can be clean water, but also at times muddy masses, which again reminds us of a mud volcano activity. The principle is the same: Upwelling fluids (gases) as the cause of earthquakes originate brittle fracture of the rock.

Small cracks can be stimulated by higher fluid pressure to develop and grow, weakening the rock. The rock eventually reaches the breaking point, and causes the earthquake. Gold writes, gas that is present in a wider region around epicentre, and that did not escape at the time of the event, continues to weaken other rocks until they also give way. This explains the usual widening area of aftershocks (Gold, 1999, p. 163).

Fig. 26: Fountain appearance.
Fountains were originated due to violent earthquake activity, through soil liquefaction and skyrocket fountains of groundwater on 5th February 1783 in Rosarn, Southern Italy.

Suess hereby provides us with an example of "the earthquake in the Mississippi area from 6thJanuary 1812, where water fountains shooting up to five metres high with loud explosions in the valley bottom below the mouth of the Ohio. Analogous phenomena were also observed by other heavy quakes, such as the one on 11thJanuary 1838 in the alluvial land of the Dimbowitza in Romania, on 12th January 1862 on the south of Lake Baikal, on 10th October 1879 in the Danube floodplains of Moldova, or on 9th November 1880 in Savetal near Zagreb (Tollmann, 1993, p. 147).

In some areas of North America, there are veritable fields of earth mounds. Individual elevations are up to 20 meters high, having a diameter of up to 100 meters. Most are however much smaller. They are often located in areas where earthquakes still occur. Most of these natural mounds are misdiagnosed because they look like Indian buildings, so-called "earth works" respectively burial mounds, such as those are exist in North America *and* Europe. But, naturally evolved earth mounds are partly used as burial sites. East of the Mississippi, there were millions of earth mounds, of which only very few were investigated, whereas millions of others were simply razed to the ground. We therefore do not know how many of these hills were naturally formed. Earthquake zones provide a hint to this problem. On 16th December 1811 residents of New Madrid, in the U.S. state of Mississippi were awakened by subterranean rumblings that sounded like a distant thunder storm. The ground vibrated and shook so violently that people had to hold themselves on to support in order to avoid falling down. The crowns of large trees snapped off and felt to the ground. Large gaps or cracks evolved from which muddy water, large chunks of blue clay, coal and sand were ejected after the quakes were over, and a rumble apparently announced a gathering of new forces for further shakes. A total of 28 earthquakes were recorded that day.

Witnesses report: In some places something came out of the Earth as air puff ... there were explosions such as by the firing of cannons that could be heard at a distance of few

*Fig. 27: **New Madrid earthquake 1811-12**. **Picture left:** a crater type "sand volcano" (arrow) in a devastated area in Pemiscot County, Missouri. **Picture right:** A typical marsh landscape that virgin evolved as a result of suddenly sinking landscape during the earthquake in New Madrid.*

miles … in the night, flashes of light sometimes seemed to come out of the Earth (Haywood, 1823). And other witnesses report a loud noise and hissing like the escape of steam from a boiler, accompanied by enormous boiling of the water in the Mississippi River in to a giant swell. Interesting are the reports of "lightning, such as experienced by a gas explosion … (and) accompanied by a complete saturation of the atmosphere with sulphur vapour". The Earth's surface rolled like waves in the sea, with visible indentation of several feet in between. These outbreaks ejected enormous quantities of water sand and coal (Fuller, 1912).

By gas eruptions that are accompanied by earthquakes, fluids move through porous rocks, caused by pressure changes. The water level in the soil is moved when gases penetrate from below. While violent earthquakes in such as in New Madrid, it was repeatedly reported that water level in wells changed and became turbid or muddy.

By this New Madrid large-earthquake, "sand bubbles" in the form of small, funnel-shaped craters filled with sand evolved over a distance of 500 kilometers, which are misinterpreted as ice wedges in "ice age regions". The eyewitness accounts of these gas eruptions confirm that crater like "earthquake spots" were formed, which are still visible today, and countless small Earth mounds also evolved. Reports of burning branches on some mounds suggest flame development, such as can be observed by methane gas sources at several locations in Asia Minor. The ancient Greeks already spoke about the flammable gas. The flashes of light coming out of the ground in New Madrid confirm this fact.

What conclusion can we learn from the "earthquake spot" phenomenon? This cannot be explained through plate tectonic scenarios! Earthquake spots are manifestly not found in places where tectonic plates overlap or collide and are far from the corresponding

Fig. 27a: New landscape forms. *A sand vent near Rowmari photographed by LaTouche and repro-duced by Oldham (1899), Plate 11. La Touche remarks in his report: "Subsequent to the ejection of sand, the surface sank down to a depth proportional to the amount of the material ejected, and several crater-like hollows were formed as the water drained back into the fissure. (Oldham, 1899 p258).* **Lower Picture:**

Fissure at Rowmari photographed by LaTouche and reproduced as Plate 10 by Oldham (1899). A row of eight still-standing thatched village huts can be seen behind the figures. In his report LaTouche notes: "At Rowmari, besides the fissures parallel to the bank of the river, which here runs nearly NE and SW, a large fissure runs to the SE at right-angles to the river bank for a distance at least 500 yards when it gets lost in a heel. It is said to run a distance of 9 miles from the river, and very likely extends further than I traced it. This fissure runs along the edge of a tract of ground, on which the village stands, rather higher than the level of the river bank, probably marking the line of an old river channel. Sand and mud have been ejected from the fissure to a depth of at least four feet. Other fissures branch off from this through the higher ground to the north, one of them passing through the huts of the villages." *(Oldham, 1899 p.258)*

structures of tectonic shifts. There are no down-welling plates or slide-slipping continental blocks nearby or continental plates shearing against one another. Earthquake spots are far from active tectonic structures. Further, the earthquake activity is limited to relatively small areas, with earthquakes around individual spots and not across widespread areas. Such "earthquake spots are best explained – indeed, can only be explained – by the upwelling-gas theory. Upwelling spurts of light hydrocarbons, especially methane, along with associated gases such as carbon dioxide, force their way up from great depths, causing fractures in the rock to open and shut repeatedly, marking the passage of these pressurized fluids. Both empirical and theoretical considerations have compelled me to draw this conclusion" (Gold, 1999, p. 157).

"What precisely should we look for? Leakage of gas driven through pore spaces by an underlying larger mass of gas will be the first evidence that can be observed. At shallow levels, this could be signalled by disturbances of the ground water and consequent changes in the electric currents this ground water carries and also by changes in the composition of the gas as components from deeper levels are brought up. Unusual noises may ensue. There may even be measurable changes of seismic velocities as pore spaces increase in number and size and the rock is thereby made more compressible. And yes, we should keep a watchful eye out for erratic behaviour in our companion animals, whose noses are more sensitive than ours, and in the animals that dwell in burrows and tunnels beneath the surface, where the composition of the gases may suddenly change and become unsuitable for supporting animal life.

Improved methods for earthquake prediction are thus one benefit we may garner from upwelling-gas theory of earthquake origins. We cannot, however, expect this theory to attract much attention among earthquake specialists until the foundational deep-Earth gas theory is more widely entertained" (ibid., p. 161 f.).

Suddenly and Unexpectedly

Let us consider the described scenarios once again but on a larger scale. From ancient times until the beginning of modern times, it was assumed that there was a water bowl or a system of water-filled cavities in the depths of the Earth. In connection with the scenario sketched in this book, the building of cavities under the Earth's crust as a result of rising water and gas is to understand that the only six to thirteen kilometres thin ocean floor can gives way abruptly.

On 1st September 1923 during the Sagami Quake, the floor of the seabed south of Tokyo dropped off 466 meters, while other areas due to the discharge were raised up to about 250 metres. The Agadir Quake off the coast of Morocco on 29th February 1960, a zone, that was over thirteen kilometres long, sunk by about 1000 metres to a depth of 1350 metres below the sea surface, while other parts were raised by about 350 metres to the water surface (Figure 28).

As documented in the scientific journal *Science*, the Central Kunlun earthquake (Tibetan plateau) of 14 November 2001 produced a nearly 400-kilometer-long earth fissure, with as much as 16 meters of a strike-slip fault in northern Tibet. The rupture length through the crust of the Earth is the largest which was originated since the recording of earthquakes started (Lin et al., 2002). The Chinese researchers discovered

Fig. 27b: The 400-kilomter long ruptures in Tibet 2001.

dislocations that are looping over mountains and valleys of the country in an almost endless curving line.

Such a long rupture can be explained in terms of local structural changes in the subsurface, but not by the breaking forces that result far off at the plate boundaries. For such a tectonic case there would have had to be many more cracks, especially in the cross direction due to the inevitably produced lateral tension. Earth cracks such as in Tibet can thus not be explained in terms of plate tectonics theory!

In this book, the same pattern of explanation has basically been offered for all geological phenomena that have been discussed up to now. Thus, in addition to "normal" earthquakes it is possible to explain middle and deep level earthquakes which occur *within* and not just on the edge of shifting tectonic plates.

Even the mystery of the only one active carbonatite volcano can be resolved. The no more than 600 degrees Celsius "cold" black lava of *Ol Doinyo Lengai* volcano in Tanzania contains almost no silicon, but carbonate minerals, a stable form of carbon. The molten carbonate "lava" is therefore formed in a separate, carbon-rich crust sphere (*Nature*, vol. 459, 2009, p. 77–80).

Another unsolved puzzle is the conductivity of the upper mantle: Although it (allegedly) consists mainly of olivine, a mineral insulating absolutely, there are naturally electric currents at depths of 70–350 kilometers. How is this possible? In laboratory tests, researchers of the *Institute des Sciences de la Terre d'Orléans* (ISTO) investigated "lava" from the carbonatite volcano in Tanzania. The scientists led by Fabrice Gaillard noted that these liquid carbonates are highly conductive. This material possess a more than one thousand times higher conductivity than basalt, which previously was considered to be the only conductive component of the mantle. Based on those liquid carbonates (lava) and it's conductive qualities, the researchers conclude that the conductivity of the upper

Fig. 28: stability. During the Niigata earthquake on 16th June 1964 (with the magnitude of 7.5 on the Richter scale) 2,000 houses were totally destroyed. In some areas occurred soil liquefaction, so that various buildings tipped completely, some nearly to an angle of 90 degrees. Quite a few flat roofs stood almost perpendicular (see arrow). Also bridge foundations toppled and some bridges collapsed. At the same time, a tsunami swept larger vessels ashore.

mantle is generally attributed to this material (*Science*, vol. 322, 28 Nov 2008, p. 1363-1365).

Moreover, in this book it is given a summery that all alleged geological evidences of the Great Ice Age can be explained totally different and first of all it is possible these different phenomena to justify consistently. As already extensively discussed in my book *Mistake Earth Science*, no "Great Ice Age" ruled over about 2.6 million years. Instead of this there was only a "catastrophic" arose and only a short time ruling "ice age" with violently snow storms and a rapid formation of ice – because without snowfalls, there *can grow no glacier*. Therefore, this short-time ice age I have named "Snow Time". During this shorter period there was tempestuously snowfall.

When the formation of alpine creeks – which flow gurgling in a river bed with rounded coarse gravel into a valley – is for example attributed to the "Ice Age" and the rounding of the gravel attributed to the long-term action of glaciers, then it is in most cases a misinterpretation.

How such alpine creeks virginal evolve from one moment to the next in express tempo could be observed on 18th May 1980 during the eruption of the volcano Mount St. Helens in the US-state Washington. It began with an earthquake of magnitude 5, followed by a *huge bulge* on the north flank of the mountain with a diameter of over one kilometer. The peak of Mount St. Helens was then blown off by an explosion. Hot gigantic pyroclastic flows of almost 300 degrees Celsius sprouted out together with turbulent liquid mudflows (lahars) at a speed of 250 kilometres per hour.

Fig. 29: Sudden filling. With the volcano eruption of Mount St. Helens in the U.S. state of Washington in 1980, this gravel field was heaped-up in express tempo to build a creek maiden-like that subsequently dreamily meandered in a curved path between the rounded debris gravel. Previously a dense forest had existed here, as can currently be seen on the left "river bank" (right hand side in the pictures). The formation of such a landscape, and the roundness of the boulders, for example in the Alps, is attributed to the long-term action of glaciers. The cause here has been proven to be due to a volcanic eruption (Zillmer, 2005 and 2007).

In a few hours new up to 50 metres thick geological layers were originated which is being caused by the up to 180 meters high mudflow (called *lahar*). More interesting for our consideration is that a thick layer of pebble stones was poured in a maiden-like created streambed, where a forest had previously existed (Figure 29). The fast a lightning formation of a flowing mountain creek between rounded rock debris occurred at a velocity of belief it or not 65 miles per hour within just 15 minutes. There was neither a creek nor any form of debris present prior to the volcanic eruption. Two different creeks evolved in such a manner, one on the northern and one on the southern slope of Mount St. Helens, shown in the German documentary film "Contra Evolution" (Zillmer, 2007).

The already named geologist Sandberg (1937, p. 47) notes, "that the evidence is until now completely missing, that the so-called 'glacial' characteristics could really have been produced as a result of glaciation". Since this time it has changed nothing - must be added. Already at that time Sandberg knew the real principles, because "the factor of flow sides had not been given much consideration. Such flowing rock streams have to take in consideration with huge water and gas masses, so that one is forced to think of a catastrophic event. But this can also occur in another form from either local or regional" (ibid., p. 49).

I want to express it more clearly: The evidence for the structural formation effect of gas and water in the Earth's crust that was gathered through the experience of the pre-

Darwinian geologists up to the beginning of the 19th century, was quite simply scientifically banned or are no longer given any consideration, since according to modern geology imperceptibly extremely slow changes pursuant to Charles Lyell have to be regarded as the determining factor, in order to satisfy the gigantic time requirement of evolutionary theory. However, fast and not extremely slow changes are determinative of the alteration of the Earth's surface texture, to which besides catastrophic water activities, especially gas or volcanic eruptions have significantly contributed.

Also the gas deposits cannot be very old. Researchers have calculated that these *would have been emptied in just a few ten thousand years* even under rocks with very low permeability (Nelson / Simmons, 1995). Can such a large gas deposits regenerate constantly through biological processes? Evidently not! This implies that gas deposits must always be filled up from depths beneath them! In the U.S. Gulf region it has already been proved, that hydrocarbon deposits always filled up (Whelan et al., 1993 and 1997). The same was demonstrated in connection with the Gulf region through local tests and experts are convinced that "hydrocarbons are continuously rising, thereby leading to an annual increase of oil reserves in this area" (Mahfoud & Beck, 1995)!

3 The Electric Solar System

The Hamburg climate model (ECHO-G model) depicts that the sun has influenced the climate on the Earth to a large extent for the past 1000 years and volcanic eruptions have resulted in cooler globally temperatures. This model together with the atmosphere-ocean model fits very well with the large number of individual reconstructions in the past 1000 years" – this statement has been made by geoscientists in their German book Klimafakten (English: Climate Facts) (Berner / Streif, 2004, p. 220 f.).

Moonquakes

The moon has been considered to be geologically "dead" respectively inactive since a very long time, but there are at least five structures on the moon which are suspected to emit carbon dioxide and even water vapour. These places are found, for example, on the intersections of the lunar rilles (Rille is German for *groove*), which are fragile points (cracks) of the lunar crust (compare Schultz et al., 2006).

In recent years, about 500 moonquakes per year were registered by seismometers. Such instrument was already equipped with the *unmanned* lunar probe Ranger 3 in 1962 and deposited on the lunar surface. Most of the quakes lie at an intensity of 2 on the Richter scale; the strongest one lies at 5. More than half of the quakes occur at a depth of 800 to 1000 kilometers. Since the moon is supposedly geologically inactive, one seeks an official explanation in the so-called tidal forces which are caused by the interaction of gravitational forces of two rotating celestial bodies and the respective centrifugal acceleration. According to this interpretation, the quakes reduce those peak values of inner tensions which are reached respectively at the points of the lunar orbit closest to the Earth and the one which is farthest from the Earth.

One would therefore expect that the moon quakes spread over the entire shell in the depths of the moon, where tidal forces should be responsible for the quake. But this is not the case. There are apparently only about 100 special locations – each with a diameter of just few kilometers – at which a majority of the earthquakes have occurred. The reason for the location-specific concentrations of the quakes can neither be explained with the help of the tidal theory nor with any other conventional theory!

Let's try an alternative explanation similar to the Earth in the sense of the gas degassing theory. "The only conceivable cause of the deep, internal quakes that have been observed on the tectonically frozen moon is the opening and closings of pores as fluids upwell. There, as with the earth, those fluids should contain primordial hydrocarbons" (Gold, 1999, p. 205).

The alpha-particle spectrometer equipped with ten detectors aboard the unmanned lunar probe *Lunar Prospector* (not Apollo!) measured the radioactive chemically elements radon and polonium, which are gradually released from the interiors of the moon. Although active volcanism should have stopped on the moon long ago, it seems that there are certain events happening where radon, nitrogen, carbon dioxide and carbon monoxide are increasingly released. At the time of quakes, an instrument placed on the moon detected gas particles of mass 16 (in atomic mass units). Thomas Gold asked, if there is "any atom or molecule of that mass that would be stable and unreactive enough to have made its way through the lunar rocks, other than methane" (ibid., p. 205).

If these fluids escape in the form of gas from the lunar surface, one should be able to notice this, because, contrary to earlier assumptions, the moon has a very thin "atmosphere", which creates a smooth transition to the interplanetary space in the form of an exosphere. The exosphere is not empty but it is filled with the (solar wind) plasma, which is quasi-neutral, but even so electrically conductive.

In fact, it is reported for centuries of short-term local brightness or colour changes on the lunar surface. This "Moon Lightning's" (Moon blinks) are described as *Lunar Transient Phenomena* (LTP) or *Transient Lunar Phenomena* (TLP), i.e. transient or short-lived occurrences on the moon. More than 1500 such sightings were reported even after the journal *Science* had already analysed almost 400 reports in the year 1967. There could be three different visual forms (Middle Hurst / Moore, 1967): transient optical phenomena at the periphery of the Mare (= dark areas), light phenomena and lightning (= Moon blinks) within the craters as also darkening's in level lands around these.

Clouds of colours white, grey and mainly red were observed, which cast partly visible shadows. In the *National Geographic Magazine* (February issue, 1972) has been reported due to the (alleged) experience of Apollo 15 on magnetic field of the moon, lunar quakes, water vapour and an extremely thin atmosphere (ibid, p. 245). A number of cone-shaped volcanoes were discovered, from which gas came out (ibid., p. 250). One saw mysterious clouds of smog and coloured lightning's in the Aristarchus crater which was considered mysterious and undefinably (ibid., p. 252).

Halo-phenomena and light clouds right above the lunar surface are signs of water clouds. These optical effects are caused by diffraction of light on the cloud droplets. This type of *photometers*, which appear reddish on the outer edge can only to be seen on very thin clouds with a uniform droplet size. Apart from these, there are other optical phenomena which occur on the Moon and are known from Earth, such as the alpine glow or phenomena similar to luminous nocturnal clouds, which are caused due to the dipole character of water molecules (so called water cluster ions) and contrary to the Earth, these phenomena are temporary phenomena on the moon.

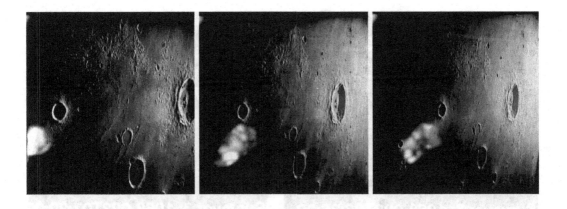

*Fig. 30: **Luminous phenomena.** The three pictures show a series of photographs supposedly taken by Apollo 14 (Photos: AS14-70-9835-9837).*

If there is water vapour on the Moon, then other optical phenomena are also possible. Such other optical phenomena can be partly recognised from the photographs of the Moon: halos, coronas, fogs and maybe even rainbow phenomena? On some photographs, we also believe to have recognised looming's which look like sun reflections of a photo lens, but these could have been caused by reflection of light at well-marked density edges, which could exist on the border between the clouds and the exosphere of the moon.

The spacecraft *Surveyor* which landed on the moon from 1966 to 1968 photographed a glow at the horizon in the nightfall. The photos also show light bands or twilight rays where sunlight was obviously filtered through the dust above the lunar surface. This occurred before each lunar sunrise and after each lunar sunset. In addition to this, the photos of unmanned probes also depict something like a hint of atmosphere at the horizon of the moon.

Previously it was thought that the Moon was a lifeless celestial body, and in the physical cosmology, there was no room for electrical interaction in the planetary system. However, the Sun emits ionized thermal plasmas, over which electrical charges can be transmitted. This plasma hits the Earth's moon, which shall have no own magnetic field, and causes an additional interaction with the Earth's magnetic field. Therefore, there must be electricity on the lunar surface. On the Earth it is different, because it is surrounded by air, a poor electrical conductor.

The ultimate source of electrical energy of the Earth is in the Magnetosphere, that is, where the magnetic field of the Earth dominates. The solar wind compresses the magnetosphere together on the dayside (= sunny side) and pulls apart it on the night side into a magnetotail. If the moon enters the tail of the magnetic field which is fluttering in

77

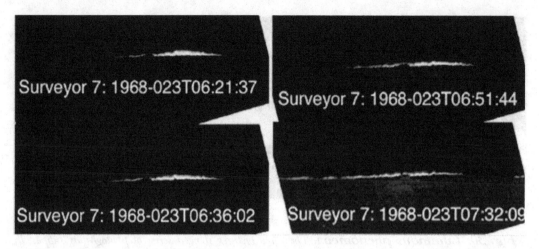

Fig. 31: Moon glow. *In 1968, the NASA probe Surveyor 7 Lander photographed "glow on the moon horizon" several times after onset of darkness.*

the solar wind, then electrical energy is induced on the moon surface, since the moon moves through the interplanetary magnetic field or the solar wind plasma.

From the sun, this plasma flows at high speed but also directly on the lunar surface. Similar occurrence takes place with the Earth at the sharp outer boundary of the magnetosphere, called magnetopause, through which an electric convection is produced: "The electric potential between dawn- and dusk side reached values of about 10 to 200 kilovolts, depending on solar activity" (Volland, 1991, p. 296).

Parallel to this a horizontal electric field should be formed between the day– and the night-side on the lunar surface. The dust particles found on the lunar surface would charge electro statically in this manner and thus would float without atmosphere. The small microscopic particles, which cannot be seen on the photos of the lunar surface, would move horizontally in the range between day– and night side as a consequence of the electric field. On photographs of the Moon's dayside taken by unmanned probes, an apparent (dust–) atmosphere is seen on the Moon's horizon and seen from the night side there could occur a kind of very gentle aurora respectively glow at sunrise.

Auroras also occur on other planets. In October 2006, these were researched more closely at the North Pole of the ringed planet Saturn. Sometimes the whole atmosphere glows to the north of 82nd latitude and then disappears within 45 minutes. Researchers say that the glow can't be explained. "If we clarify the source of these phenomena, we will undoubtedly discover a new physics, which will be applicable only to this unique environment" (Stallard et al., 2008). Currently, the auroras on Earth, Saturn and Jupiter are each explained by a different cause. One should however, be able to explain these phenomena through an all-encompassing theory or cause!

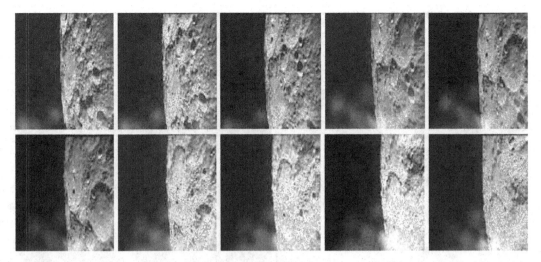

Fig. 32a: Moon clouds. *This degassing process was photographed during Apollo 10 mission. Pictures: AS10–33–4955 to 4964 (NASA).*

Back to the Moon. After the data of the lunar probe *Lunar Prospector* from 1998 to 1999 was prepared. it was only recently published in February 2007, that the surface of the Earth's satellite is electro statically charged by the Earth's magnetic field and solar wind. The surface charges over large areas correspond to voltages of up to 4500 volts. This leads to accumulation of electrically charged dust. This also leads eventually to the endangering of all types of metal or metal devices through short circuits in projected manned lunar missions (Halekas, 2007). But what about the Apollo Lunar Modules?

After all, the *Lunar Prospector* was the first lunar probe, 30 years after the (often doubted) Apollo missions. Although it was believed that one knows everything about the "dead" moon, one now wonders about the regional emergence of strong magnetic fields with diameters of a few hundred kilometres, which build around them a type of magnetosphere similar to those of planets (Lin, 1998).

Let us now briefly discuss new results sent to Earth by the Messenger Spacecraft during its path across Mercury in 2008, which has been considered an equally inactive celestial body till now.

Totally Different Mercury

Robert Strom (*University of Arizona*) confirmed that the researchers now view planet Mercury as totally different from how they viewed it 30 years ago. The largest surprise in the new Messenger photographs is a strange structure of the middle of the Caloris Basin,

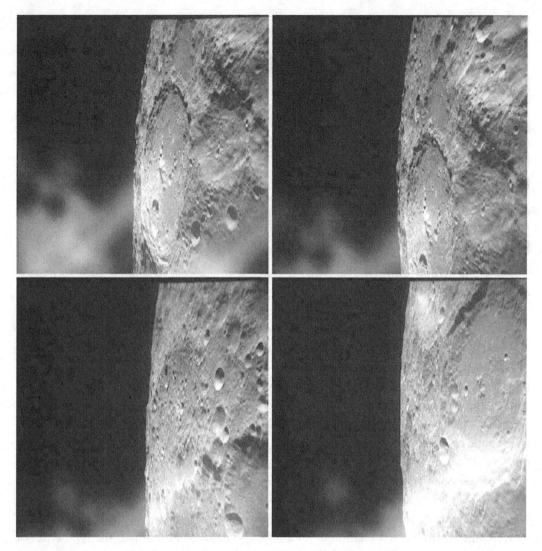

Fig. 32b: Active moon. *Photographed supposedly by Apollo 10, gas emission process on the moon and lightly colouring areas. Photos: AS10-32-4790 to 4795 (NASA).*

named by the researchers as *Spider*. More than a hundred flat trenches move like rays away from one large crater, which is 40 kilometers wide. The researchers suspect that these rilles are expansion cracks. The emergence of these remains is a mystery.

Let us assume that on Mercury like on Earth, methane and other hydrocarbons flow upward through crack zones in the rocky crust. Because – without any biological process – oxygen is present in the thin mercurial atmosphere to an extent of 42 percent, which is comparable to the Earth's exosphere, and oxygen is also present in the soil. Therefore an oxidation of potentially upwelling methane can occur. With this abiogenic process arises carbon dioxide and water. But both kind of molecules do not seem to be available in Mercury's atmosphere currently.

Stable liquid water is also not available on Mercury, but the reason could be that it could have evaporated as a result of the lower pressure. The light water molecules escape either in the planetary space or they are decomposed by solar radiation into hydrogen and oxygen, which enrich the oxygen component of the atmosphere. As against this, the light hydrogen should have escaped from the thin atmosphere into interplanetary space, as the "gravity" is not strong enough to hold back. In fact, in the Mercury atmosphere, in addition to oxygen, there is about 22 percent of hydrogen currently, which is continuously renewed through an ongoing production process due to its "volatility".

However, a portion of the hydrogen can interact directly with the soil material. This is eventually also applicable to the water produced by the oxidation of methane, which may not necessarily be melted because of the low atmospheric pressure, but it could directly pass over into the gaseous phase, without running through the melting phase. This explanation could be applicable even for small ice formation on the Earth's moon, or on comets. Contrary to this statement, it is believed that ice was introduced by comets – because water that could freeze, it is supposed not to occur on those celestial bodies.

Along with water, carbon dioxide is also produced with the oxidation process of methane. This carbon dioxide may lie on the surface of the Mercury after escaping from the crust or it can interact with the water molecules formed before these break down into oxygen and hydrogen. Carbon dioxide reacts easily with the calcium oxide present in the rock and forms carbonate but it is also readily soluble in water depending on the temperature. So carbonic acid could be created as a reaction product. The salts of this acid are called carbonates or hydrogen carbonates.

Since the atmosphere of Mercury, however, consists of 29 percent sodium, *sodium carbonate* can also be formed as a salt of carbonic acid. Since its melting point is 851 degrees Celsius, this salt – which is also known as soda – would be stable on the surface of mercury. Therefore *sodium carbonate* could be present around a crater after a carbon dioxide emission on the surrounding mercury surface. This process is followed by the chemical industry for the production of bleaching agents. The release of aggressive acting atomic oxygen forms the basis of the bleaching process. So the puzzle can be solved as to why the Caloris Basin (with trenches arranged in a ray-like manner) is brighter than the surroundings?

On the other hand, the grey colour reminds me of carbonate cement. The meteorite ALH84001 originating supposedly from Mars showed carbonate cement along with magnetite and iron sulphide, which exists in the crust *as a filling in crevices and in petroleum-containing layers*. These specific properties differentiate carbonate cements from the mass of the remaining carbonate rocks in ocean.

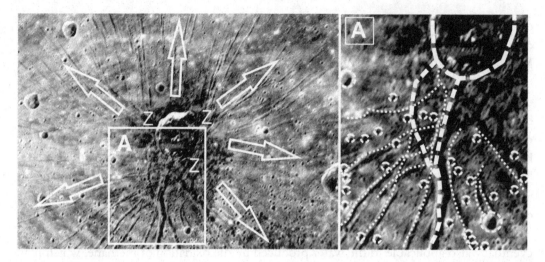

Figure 33: Mercury mystery. *Like the Earth's moon, Mercury shows craters, like the Caloris Basin here, with partly radial "rays", which are emitted from craters or centers (Z) located just next to them in the form of a star. Were the gas emission processes associated with electrical discharges?* **Right picture:** *In the enlarged picture a number of trenches have been highlighted, next to circular craters, which should be "pockmarks" caused by degassing gas.*

The cause of formation of the Caloris Basin may have different reasons wherein the velocity of the gas transport plays an important role. We know too little about the minerals in the soil of Mercury. Depending on their nature, different chemical reactions are possible; however, their origin can be traced largely in the degassing of hydrocarbons (mainly methane). The spiderweb-like Caloris Basin therefore does not represent an impact crater, but it represents a gas-emission crater, where electrical discharges are responsible for the formation of shallow trenches, whereas other "cracks" result from tensions caused by gas eruptions in the crust of Mercury.

Methane is a component of the planets in our solar system, as long as it's not too hot. Carbon monoxide and hydrogen were a component of the solar nebula. So the assumption made by the astronomers of Sara Seager – *Massachusetts Institute of Technology* (MIT) in Cambridge – is that in the other solar systems, planets could have been constituted *completely* of graphite or carbon monoxide. (*Sterne und Weltraum*, Online, 2nd October 2007). Calculations indicate that carbon monoxide would be the main form of carbon at high temperatures; and at "normal" pressure it would transform to methane at less than 326 degrees Celsius (600 Kelvin) (Anders et al. 1973).

Already, the American chemist Harold C. Urey, who was awarded the Nobel Prize in 1934 for the discovery of heavy hydrogen, worked on the problem of the formation of solids in the presence of small-grained catalysts. He claimed already in 1953 that tar compounds are the main source for carbons on terrestrial planets.

Anders et. al. (1973) has demonstrated that the molecular arrangement, which is present in petroleum, follows exactly the same special pattern, which one gets from cataclysmic reactions of carbon monoxide and hydrogen. "One can therefore speculate that such reactions took place in the solar nebula, and were in fact responsible for producing the solid compounds of carbon that were incorporated both in the meteorites and in the forming Earth" (Gold, 1987, p. 34.).

For a given mean ratio of carbon and hydrogen, and in presence of a particular set of catalytic surfaces, a definite statistical distribution of the molecular species would result. Unless the hydrogen content is very low and in the 100–300 km depth range carbon molecules are greatly stabilized by high pressure, methane is likely to be the dominant member among the mobile fraction. A lot of carbon in immobile form would also be produced. Present estimates are that methane is largely stable at these temperatures and pressures in such depths, though it may partly dissociate into carbon and hydrogen. Many heavier hydrocarbon molecules are also stable. It is methane, with an admixture of heavier hydrocarbon molecules, which would be given off by the material in these circumstances, and it is this mix that would start its upward migration towards the surface. If carbonates are present in the original mix, then it is possible that carbon dioxide also would be generated (ibid., p. 36).

In other words, carbon may be present in the interior of planets, moons and comets out of which methane may be formed. With the oxidation of methane in the depth of Earth in addition to water arises carbon dioxide, which itself is a starting material for various chemical compounds such as carbonates and carbonate cements.

Hydrocarbons are chemically formed in the depths of the Earth and these can be detected also on other planets and moons, even in absence of any biological processes (photosynthesis). In other words, although hydrocarbons are the common form of carbon on planets, moons and comets it is claimed that none of the hydrocarbons found so abundantly on the Earth originate from a similar non-biological (abiogenic) source. *Therefore the Earth would be absolutely a special case.* Why methane and ethane on Earth do not occur without any biogenic process like on the Saturn's moon Titan or carbon dioxide like on Mars?

In fact, there is a quantity issue if carbon dioxide is of biological origin, because carbon exists not only in long-lasting debris like in calcium carbonate (limestone), calcium-magnesium carbonate (Dolomite), sodium carbon or calcite, but also in sediments and in the crystalline continental shelf. It is also present in various forms of non-oxidized carbon such as petroleum, coal, tar, graphite or methane. Terrestrial carbon in this huge amount cannot originates only from atmospheric carbon produced through photosynthesis over a carbon cycle; it derives to a large extent from another source. "While hydrocarbons are the most common form of carbon among the planets, we would have to argue that none of the plentiful hydrocarbons on the Earth are from a similar source. The Earth, we would have to say, derived its large supply of carbon an unknown source material that produced carbon dioxide, a material that is not prominently represented in meteorites or expected from cosmos-chemical information. The Earth then produced its complement of hydrocarbons from this by a process unique to this planet, namely the biological process of photosynthesis. It does not sound like a very good case" (ibid., p. 36–37).

On Earth, there is a general systematic relationship between oil and gas fields and the carbonate mineral calcite – the most stable polymorph of calcium carbonate – to be find above oil and natural gas deposits in sedimentary layers. This vertically distributed arrangement can be explained by the methane that streams upward from the depths of the Earth. Similarly we can explain the phenomenon of calcite-filled cracks in the crystalline granitic bedrocks which lie beneath the sediments. As calcite is widespread in large areas of the granite basement in Canada, Scandinavia and Siberia, this points to the upwelling of methane. Can we draw a conclusion on a common effective principle of upward streaming methane from the depths of the Earth? Thus, the extensive and deeply staggered spreading of calcite (calcite) in the continental shelves can be explained. The mineral calcite is very common and it belongs to the mineral class of the anhydrous carbonates. In other words, calcite can be formed quite easily, since it is a (calcium) salt of carbonic acid, which in turn, is a reaction product of carbon dioxide with water. Both substances are formed automatically as oxidation products of methane, existing in a fluid (liquid) and even not gaseous form in the depths of Earth if enough oxygen is present. In other words, during the migration and diffusion of the light methane molecules in the direction to the surface, calcite is formed automatically in the absence of any other external processes, catalysts or other reactions.

In fact, with very little effort living creatures in warm waters generate shells, a material consisting of calcium and (chemically seen) calcite. This fact is exploited, and claimed that by far the largest calcite deposits are due to marine deposits that are formed from skeletons and shells of countless small marine animals on the seabed. Only a small portion shall to be formed inorganically namely as millimeters sized beads in warm tropical shallow seas.

Because calcite (feldspar) is present in unimaginably vast quantities and is also a constituent of metamorphic rocks such as marble or sedimentary rocks (for example limestone respectively calcium carbonate), it indicates that unimaginably thick layer of dead animals would have been deposited. But then in former times there could have been only shallow seas, because below a sea depth of 3500 meters calcite dissolves completely in water...

In any way, the calcite found frequently in the crust of the Earth could not have been formed only from the atmospheric carbon dioxide as a result of the carbon cycle or by

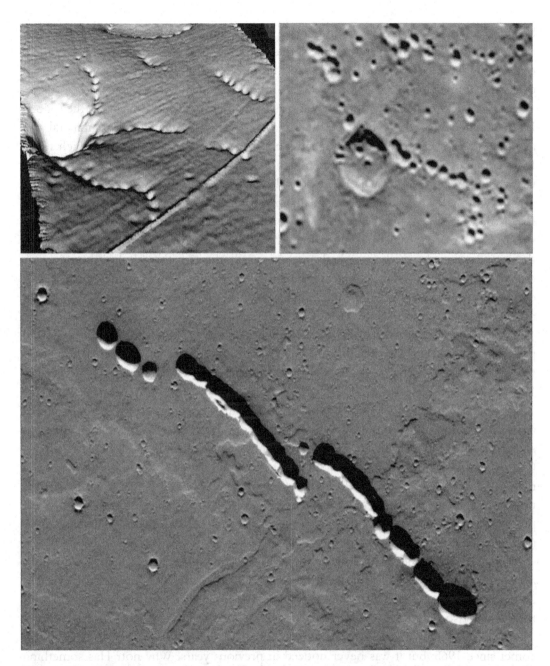

Figure 34: Planetary pockmarks. Left picture above: Pockmarks called craters at the bottom of the North Sea, to be seen in rows in the form of chains partly starting at larger craters. In the foreground, there is a 65 cm thick pipeline. Right picture above: Line style arranged craters on the Mercury similar to the craters in the North Sea. Lower figure: Craters on Mars which are chained to each other like a string of pearls. It is important to note that there are many small circular pockmarks.

biogenic processes. What *gigantic* amounts of small animals and organisms would have once been dead? The problem of the "suddenly" emerging calcium was already pointed out in my book "Darwin's Mistake". and the puzzle seems to find its solution in the degassing of methane as a reaction product of the components of the solar "nebula".

Let's go back to the planet Mercury, and ask, whether the light colouring in the Caloris Basin could have actually been calcite (calcspar)? This frequently found mineral can assume different crystalline forms but it can also assume aggregate forms, which range from colourless to milky white and rich grey. By adding other elements, colours such as yellow, blue, red, green, brown or black are also possible.

I have been offered several new explanations for the ominous light areas in the Caloris Basin for which there exists currently no conventional explanation. The mystery of the small craters, placed inside the light coloured basins areas, can be solved with the help of the gas emission theory. Escape hydrocarbons easily through these holes, then remains carbon monoxide (= carbon black) in case of an incomplete oxidation and this is deposited around the crater rim. Accordingly, the main crater would be an explosion and not an impact crater.

The Mercury is geochemically active and not "dead"! Therefore, there are optical phenomena similar to Earth's moon. The German book "Astronomie, die uns angeht" reports about "the temporary darkening on certain Mercury landscapes" (Herrmann, 1974, p. 74 f.).

A high proportion of sodium is unusual in the thin atmosphere of Mercury. Its atoms are continuously driven through the solar wind from the surface of the planet, although sodium is not accounted as the main component of the minerals on Mercury's surface. Can one explain in this manner the origin of the innumerable sodium atoms with a 2.5 million kilometres long tail? The length of the orange coloured tail is more than 1000 times the radius of the planet. This new knowledge, which is surprising for the professionals, is due to the fly-by of the spacecraft *Messenger* on 14 January 2008 (Baumgardner, 2008).

What does the planet Mercury have in common with comets? Even a large comet has a long sodium tail. To the surprise of the researchers, in early 2007, it was discovered in 2006 that the small comet McNaught (C/2006 P1) has also a tail, which lights up with the characteristically yellow-orange light of sodium.

However, carbon was also found in the tail of this comet, which was the brightest comet since 1965; but it was never noticed in previous years. Why not? Has something changed suddenly?

With the help of the NTT Telescope *(New Technology Telescope)* in Chile, it was possible to make photographs of three several thousand kilometers long spiral-shaped gas blasts. One believes, the ice from the comet has been heated and suddenly evaporated by the Sun's heat. Therefore we take a closer look at the comets, known as "dirty snowballs".

The Myth of Dirty Snowballs

The bright Halley's Comet comes every 76 years near the Sun. During its approach in 1986, this comet was investigated by five spacecraft's. ESA's Giotto spacecraft managed to directly observe its core, which is seen as an irregular shape with dimensions of approximate 15.3 x 7.2 x 7.2 kilometers. That volume is only about 420 cubic kilometers, and violates the theory that comets originated since the beginnings of the Solar System. Why? Because simple calculations show that all comet cores are too small to satisfy the "dirty snowballs" principle. For comet Halley, it was reported that there was a loss of 50 tons of comet nucleus material per second in the Sun's atmosphere. If this loss rate is constant, then this comet returns near to the Sun at most 400-times and mathematically will exist at least not any more in 30 400 years

Apparently all the comets shall be created before about 4600 million years ago. But the life duration of all active comets is due to the low core mass *maximally* in the range of some hundred thousand years. This anomaly is exclusively owing to the conventional notion of "dirty snowballs". According to this convention, snowballs shall be melt due to the heat radiated from the Sun. To solve this puzzle one came up with the idea that so called long-period comets are waiting at the edge of the solar system before getting activated one by one in some kind of a relay race.

For the short-period comets, which are supposed to come from the Kuiper Belt, which shall be located behind the orbit of Neptune, one believes that there are collisions among objects in what way occur fragments which are bounced into the solar system. A purely conceptual model – totally unproven! Of the 70,000 presumed Kuiper Belt objects with a diameter of more than 100 kilometres, only 800 have been discovered until now.

While about one percent or in other words *only at least* every hundredth of the presumed Kuiper Belt objects has been discovered, the after Jan Hendrik Oort (1900-1992) named Oort Cloud is pure fiction, a hypothesized spherical cloud of comets, supposed at a distance of 50,000 times the distance Sun-Earth far out in space enveloping our planetary system. This spherical cloud beyond Pluto's orbit shall be the original location of long-lived comets with an orbital period of more than 200 years. Neither the suspected comet cloud was discovered, nor has been submitted a serious conceptual model for a scattering process of long-period comets. It is speculative, to declare the influence by passing stars or a not yet discovered planets as the cause.

Why do comets exist at all and how do the short-lived active "dirty snowballs" break apart during the lifetime of human beings, like the comet Biela in 1845 – few seconds before 12 o'clock of the cosmic time clock? The short-living comet *Schwassmann-Wachmann-3* began to break apart in 1995 and during this process; it became a sensational 250 times brighter than normal. A strange snowball! In the spring of 2006, the comet

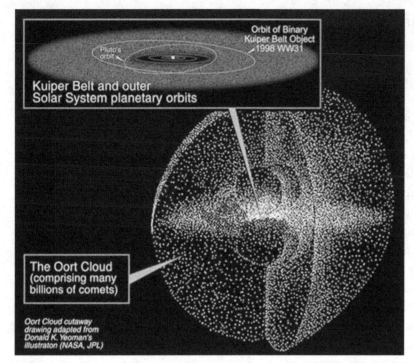

Figure 35: pure fiction.

Short-period comets should be in the disk-shaped Kuiper Belt and long-periodic comets are believed in the hypothesized spherical Oort Cloud.

broke apart which was noticeable due to an increasing brightness! During its maximum closeness to Earth on 13th May 2006, both big fragments reached a greater brightness than the original one at the time of the first disrupting in 1995. Also, the form of one fragment changed several times in the starting of May 2006. There were strong brightness outbursts. An interesting question: Why does this "dirty snowball" become brighter and brighter, when it breaks apart and why do the parts shine so long time after they have broken apart?

Conventional astronomy is based on regularities that describe a mechanically functioning universe with electrically *neutral* bodies in a vacuum affected by mass attraction respectively gravitation also taking into account the theory of relativity – without electrical effects as a fundamental principle. Most astronomers are confident that our solar system has emerged from a gas and dust nebula through purely mechanical interactions of its particles and gravitational forces. In this theory called as the Kant Laplace's theory, the electrostatic and magnetic forces play no role in the emergence and existence of planetary celestial bodies.

The apparent short-life of comets is a "birth defect" of this Kant-Laplace theory. The alternative would be require a much younger solar System – a thought model that has originated only due to a wrong theory. We rethink the situation and reconsider the latest research results that more than 99 percent of the matter in the universe consists of plasma, which is electrically quasi-neutral, but it contains particles that are electrically charged. Therefore, the plasma can transfer electric forces between solar systems as also from the sun to the planets.

Black Nucleus

Consider first the nucleus of comet Halley. Not only professionals were totally surprised when the spacecraft reported while approaching the comet nucleus in 1986, that the surface of the "Snowball" is pitch black (albedo is 0.05), covered by hydrocarbons or carbon. This substance – misleadingly named as "organic" – is also emitted, originated from the comet nucleus itself. Is this black "snowball" an exception? No! There is *even a little blacker* short-period comet Wild 2 which was discovered on 2nd January 2004 by NASA spacecraft Stardust.

Images captured on 22 September 2001 by the American space probe *Deep Space 1* show that even the short-period comet *19P–Borelly* has a core, which is dark like black toner powder (albedo 0.02). Some spots are even so black that a reflectance value of only 0.007 was recorded for which hardly any corresponding minerals are currently known. From dirty "snow balls" they have become coal-black ones.

A question is not going to be discussed: Why are these comet nucleus apparently covered with carbonaceous material, which remind of coal dust or soot? Dirty snowballs should be different. Meanwhile, the conventional view of comets was modified, and the dirty ice is supposed to be hidden beneath the black surface. These unproven intellectual game only serve the purpose to perpetuate the old (out-dated) theory, otherwise a completely new cosmology would have to be established.

Now let us look closely at the comet's tail. Previously, one was convinced that the tail is of material similar to the core and consists mainly of water molecules, as expected with a snowball. Until now, scientists believe that the tail matter consists mainly of water molecules related to the core, as expected due to a snowball. But in contrast to the conventional comet theory, the tail is not wet, but relatively dry. A team of researchers at *Arizona State University* discovered through telescopes during the study of comet Hale-Bopp Hyakuta that the proportion of ionized carbon monoxide molecules are predominate at a distance of ten to twenty million kilometers away from the comet's core. Celsius. From these molecules can be formed methane at a temperature below about 326 degrees Celsius.

Previously announced were two different types of tail: a so-called dust tail, which follows roughly the orbit of the comet, and a gas tail, which shows almost exactly away from the sun. To the total surprise of the experts behind Hale-Bopp was discovered a third tail, which consists of sodium. The length of this tail was 50, according to some estimates as much as 100 million kilometers. This sodium tail was situated between both others (cf. Cremonese, 1997).

A convincing explanation for the origin of the third tail does not exist, because the lifetime of sodium atoms in a solar distance of one astronomical Unit (= distance of Earth from the Sun) with just under 14 hours is timely too short to originate such a long tail. This sodium has also not originated from the comet nucleus. Are these electrochemical processes which have their origin in the solar wind plasma? As the speed of the sodium atoms is too high, so this cannot solely have caused by the sunlight. This means that energy input must have been resulted from the interplanetary plasma, kinetic energy generating.

Thus the origin of sodium is not yet clarified. It is indeed not a specific problem of comets, since even the planet Mercury has a sodium-tail. And the matter is further complicated because Jupiter's moon Io also emits a cloud of atomic sodium. The orbit of Io passes in a tubular plasma torus, which is extended in the hydrogen corona around Jupiter and rotates rigidly around its axis with the rotation of the planet. After the discovery of the Io plasma torus with significant mass density, however, electric currents which are acting vertically to the magnetic field can no longer be neglected (Neubauer, 1991, p. 198). In dense regions of the Io atmosphere existing electric currents leave these loaded with plasma and establish a sort of "comet tail" (ibid., p. 200).

Jupiter has the most intense radiation belts in the Solar System. Electric currents are responsible for ensuring that a Jupiter surrounding plasma torus orbits rigidly with the rotation rate of the planet. This effect is called synchronous rotation and means that one side of Io always faces Jupiter analogical to our Moon and the Earth. From the surface of the Io, Jupiter appears to be locked in place in the sky as it slowly rotates. Overall, starting from Jupiter an electric field is radially directed into interplanetary space (Fig.36). This is the strongest energy alteration of planet-satellite interactions by far in our solar system (ibid., p. 195 ff.)

With electrical interactions or absorption of energy by the plasma, we are also able to give an answer to the puzzle discussed up to now, why Jupiter radiates much more heat as getting from the Sun. Conventionally, scientists believes in the so-called Kelvin-Helmholtz effect: Due to release of heat into the ice cold space, there shall be a cooling and subsequently a compression of the planet takes place. Through this process should be released additional internal heat. In other words, by releasing heat into space there is believed to create so much heat in the gas planet themselves that twice of heat is totally emitted as received from the Sun. Comment unnecessary…

On Jupiter, there are new research findings of a 70-year climate cycle. In such a time period, several hurricanes are formed that disintegrate after a certain time. Even in terrestrial hurricanes, electric fields have been measured recently. Could it be that

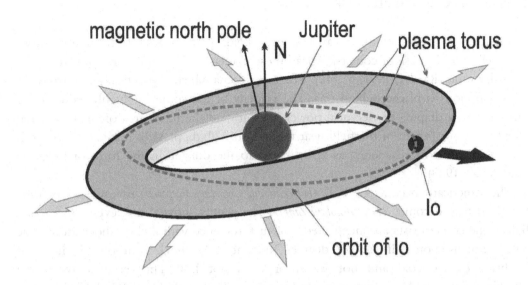

Figure 36: Plasma-torus. *The orbit of Jupiter's moon Io proceeds in a (tubular) plasma torus. Electric currents run through the dense regions of the Io atmosphere loaded with plasma and they form a sort of "comet tail". Io gets an upstream flow of plasma, which is synchronously rotating with Jupiter inside the plasma torus. In this way, with a typical magnetic field will be originated an electric field that is radically directed in an outward direction, schematically indicated by arrows. According to Neubauer, 1991, p. 192.*

electric fields are the original source of the formation of hurricanes and not causally the temperature in the atmosphere?

At the end of 2008 a study – published by a Mexican-Bulgarian team – provides a statistical correlation between the occurrence of hurricanes and cosmic rays or solar activity. The intensity of cosmic radiation rose strongly few days before the onset of the hurricane. This affects the formation of clouds in the atmosphere. Therefore not only for this relationship is discussed a possible correlation between so-called "space weather" and the terrestrial weather and climate (Pérez-Peraza et al., 2008).

Does the climate on Earth also follow a solar cycle? The climate model (the ECHO-G Model developed at the University Hamburg, Germany) – proposed for the first time in 2003 – makes clear that the Sun has influenced the Earth's climate to a large extent for the past 1000 years and the volcanic eruptions have contributed to short periods of cooler temperatures... The Hamburger ECHO-G results for the time period 1948 to 1990 clearly indicate the strong influence of solar variations on the earthly climate history, German geoscientist's confirm (Berner / Strip, 2004, p. 220 f.).

Electrical Gas Discharges

The formation of sodium tails and electrical fields at Io suggest electrical interactions – as well as in the case of comets? Ahead these comets, a bow shock are formed against the solar wind (Raeder, 1991, p. 342). The inner coma which is practically free from the solar wind ions is placed behind this. The velocity of the comets is highly reduced and this leads to a draping of electrical power lines around the inner area of the coma, and "for the strong growth of the field near the nucleus" (ibid., p. 349). Inside this system is converted much more energy, as is delivered to the outside in the form of radiation (Gary et al., 1988).

The American space probe ICE (*International Cometary Explorer*) investigated in 1985 the short-period comet *21P/Giacobini-Zinner*. The results showed that even 188,000 km before the comet exists an unexpected violent turbulence with a clear shock front. The temperature is about half a million degrees. In front of the shock front towards the sun it is, however, ice cold and not warm or even not hot! The temperature in the interplanetary space between the sun and the comet is only about –270.5 degrees Celsius (2.7 degrees Kelvin).

It is clear that heat is generated primal in the bow shock of comets by the conversion of the flowing energy of the solar wind. Pure heat radiation cannot produce such temperatures and luminous phenomena of comets, but still, the energy has to come from

Figure 37: Luminous points. *The Esa probe Giotto produced photos from the nucleus of Halley's Comet at first time in 1986. The comet nucleus does not light up completely. There are light phenomena only at certain points where probably superconducting substances are located. These are the sources of the cathode rays directed towards the sun, which turn over at a certain distance, revolve around the comet head and form a tail (left picture: after Vollmer, 1989). Right picture: © Esa.*

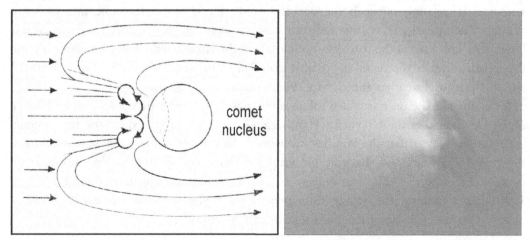

the Sun. But how is it transferred? After the Kant-Laplace theory the interplanetary and interstellar space will contain only isolated dust particles and gas molecules in a very thin distribution. The old idea of "ether" that fills the entire intergalactic space, is rejected nowadays, probably rightly. If "ether" exists, then this raises the much-debated question of whether and how energy, heat and light can be transferred through absolute nothingness respectively a vacuum.

Now, gas is very sparsely distributed throughout the universe. Even at very high voltages, no current flow is detected. However, if the gas is heated to high temperatures then plasma comes into being, whereby the attributes change hugely – this

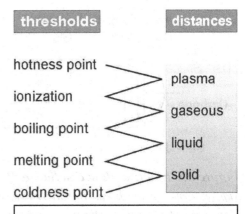

Figure 38: Plasma, the fourth aggregate state. More than 99 percent of the visible material in the universe is in the plasma state, thus in form of a (partially) ionized gas.

is called as the fourth aggregate state. By heating the gas molecules are ionized, thus the formed plasma is a conductor of electricity. As whole, however – in not too small parts and not at too short times – this plasma is electrically neutral to the outside (in a stationary state).

More than 99 percent of visible material in the universe is to be arranged in the plasma state, which is resistant to small interruptions. The stability leads back to the far-reaching electric power, what for a pure "gravitational effect" has not the ability. Each charged particle is in interaction with others, also those that are distant. This means that through the plasma is flowing electricity between the Sun and other celestial bodies in our solar system. Thus, everything which is electrically conductive is quasi connected respectively communicate with each other.

A working hypothesis can be formulated (see Vollmer, 1989): The Sun acts as an anode (positive pole) and from this, electrical energy flows through the plasma towards the cathode (negative pole), which in our case forms the comet or some other electrically conductive celestial body. Now we can easily explain why a comet begins emitting bright light only at a certain distance from the sun. For the *ignition* of the comet a certain voltage, the ignition voltage, is required. This is achieved when on the base of a small existing "pre-conducting current" enough electrical energy is transported from the Sun (or a different source) to the comet at a certain distance from the power source (Sun). Thus a regeneration of charge carriers begins. The electricity is rapidly growing like an avalanche, and the result is an independent plasma discharge, why bright light is emitted.

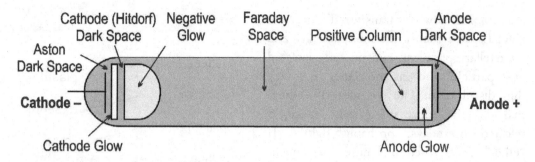

Figure 39: Electric glow discharge. *If one looks in analogy of the theory of electricity, then the Sun represents the anode and electrically conductive celestial bodies within the reach of the solar wind plasma constitute partly a cathode. At a certain distance the ignition temperature is reached and comets start to glow. The brightness distribution of the glow discharge tube is corresponding to that in the solar system: the cathode should be a comet and the anode is the Sun. The Faraday space (= interplanetary space) is relatively dark between cathode and anode although both (comet and Sun) are brightly.*

Technically, it is an electrical glow discharge in low pressure plasma which represents non-thermal plasma because of the significantly low pressure in the vacuum. This glow discharge indicates a spatial structure which is directed from the cathode to the anode in a glow discharge tube filled with hydrogen or inert gas. Closest to the cathode is located a thin dark space, called the Aston dark space. Then follows a thin light skin (the cathode glow), which is the first cathode layer. Next there is a faint zone, which is also called as a cathode or Hitdorf dark space, followed by the brightest part of the discharge process, the negative glow. In front of this light phenomenon begins a lightless zone, which represents the interplanetary space in our solar system. This dark zone extends up to the Sun and is called Faraday (dark) space in the electrical glow discharge tube. Analogous to the light phenomena cathode glow and negative glow at the cathode (comet) before the anode respectively Sun there is positioned a bright positive column and an anode glow.

Can one recognise the layered structure of the glow discharge in front of a comet nucleus in the universe? As the very bright comet Donati went through the nearest Sun point (perihel) in September 1858, its development could be closely monitored by large telescopes. On 23rd September the telescopic sight was extraordinary (Weiß, 7th edition, 1886, p. 81):

"The core, only 2100 kilometres in diameter … now had … its greatest glory. Around the nucleus there is a clear bright shell, whose apex is directed toward the sun, and it extends up to 10,000 kilometres from the core. The outline of this light phenomenon cover is dark. The boundary of a second, less bright shell is around 20,000 miles from the core at its peak. This second cover is also framed by a dark bow, outside which there is a halo of fine scattered nebulosity that decreases rapidly... "

This description is identical to the described discharge glow in an electric discharge tube. We are dealing with a permanent independently running gas discharge at low gas pressure. The resulting light emissions through appropriate discharge are used by us technically in fluorescent tubes and glow lamps. The conformity with the cosmic phenomena emerges very clearly at comets because one can conclude from the colour of the phenomena about its linkage to a cathode or to an anode. The colours of the stratified column or the negative glow light are in characteristic colours depending on the type of gas and one can determine whether nitrogen, hydrogen, helium or sodium causes the gas lighting. Based on this conformity of light phenomena in the electric glow discharge tube (vacuum tube) and in the planetary system, there is no doubt in the anode-cathode principle.

In the universe there is an even more ideal vacuum than in a discharge tube (gas-filled tube) in the laboratory and moreover the temperature in the universe is much cooler at only 2.7 Kelvin or about –270.5 degrees Celsius. Therefore, in contrast to the discharge tube, a comet could act as a superconductor, if this comet consists of certain minerals and metals, at least in part. Due to the superconductivity, certain substances and also a number of chemical compounds lose their electrical resistance almost completely. Because the present space temperature is below the so-called *transition temperature* from which starts the superconductivity, the electrical resistance of such a comet, consisting of suitable materials, would drop to level at which initiated electricity can last with undiminished strength for a long time.

Such could be explained that comets are much longer existent than it would be possibly in accordance with the "dirty snowball" theory. The solar system as such becomes theoretically older because comets due to electric lighting phenomena are much more stable and are therefore older, than was calculated by the conventional snowball theory. Also, countless objects in a in a standby position in the Kuiper Belt or in the Oort cloud would not be required because comets are much older than assumed so far!

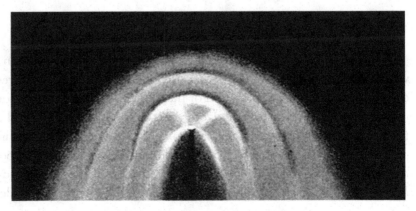

Figure 40: comet Donati. This drawing which has originated in 1858 shows different alternating bright and dark zones in front of the comet's head. Source: Weiß, 1886, Plate VII).

Figure 41: Comet Wild 2. *The shallow hollows with steep rims do not represent impact craters but on the one hand those can be areas where superconductive materials are burnt out or on the other hand such craters were caused by flowing gas jets. At least ten gas jets were active in January 2004 when the Stardust spacecraft flew across the comet Wild 2 (artist's impression of the jets).*

Although the comet nucleus itself is cold like the cathode in the discharge tube, it comes to a sharp rise in temperature before the cathode, will say, outside of the comet in direction to the Sun. At the comet nucleus heat developments occur only selectively at special points where radiate light. Cavities arise on these sites, as shown in the images of the comet Halley by the Giotto spacecraft (Fig. 37, p. 92). If this process is ongoing then a comet nucleus can break apart at one time or another.

A comet can end up as an asteroid, namely, when the comet nucleus loses its electrical conductivity at al. If the cathode particles, especially superconducting materials, are burned out, then what possible remains is a pure stone meteorite. Thus, it is also to explain that the comet *81P/Wild*, shortly called *Wild 2*, from whose tail the Stardust spacecraft in January 2006 brought samples back to Earth, has a similar composition to a typical asteroid (Ishii, 2008).

Thus, the question has been answered, why not all heavenly bodies build tails near the sun. To be the crucial factor is the electric conductivity.

The Mystery Luminosity

In a rather simple way, we have described electric interactions in our solar system, which in the context of comets can also be experimentally observed in the electrical discharge tube. For the explanation of these phenomena we should consider the gas that is everywhere in the universe, however, it is in the ionized form and hence, conducting form in the fourth physical state (plasma). Since almost all the matter in the universe is in plasma state, in astronomical explanations electric and magnetic modes of actions and reactions effects should be taken into account almost mostly. Already in the old book *Littrow, Wunder des Himmels* by Edmund Weiß (1886) it was concluded on the basis of certain observations that the cause of comet effects is caused by an electric mode of operation.

Even if today in scientifically explanations is included an ionization of atoms and molecules before the comet nucleus, accompanied an increasing of the coma; and also the elongated tail is regarded as being of plasma, reveals the dilemma of astronomy:

Plasma physics has definitely right solutions for different astronomical riddles, which could not be resolved with conventional interpretations and *can partly sufficient describe well the complex–valued effects of magnet hydrodynamics. Plasma physics is therefore the tool for understanding the processes in the universe*, but by classical astronomers it will be considered only in so far as it cannot be avoided. In other words, astronomers have been accepted electrical interactions – such as between Jupiter and its satellite Io – but this is regarded as an isolated phenomenon like the processes on comet or even in polar lights or thunderbolts on the Earth, which also plasma phenomena represent.

If one integrates the findings of plasma physics with all its consequences into prevailing cosmology, then would the explanations of conventional astronomy – which generally are based on the effects of dust and gravity (Kant-Laplace theory) – quickly become fundamentally wrong. Mechanical or classical-physical explanations are therefore generally not appropriate, although can be calculated almost exactly planetary orbits and gravitational effects with empirically developed formulas such as well. If something can be calculated with an empirical formula that says nothing about the cause for example of gravity! These old observations not take into account electrical interaction across the plasma. Newton, Kepler and other early astronomers naturally didn't know it; still their theories are almost prevailing to this day.

Of course one cannot deny that dust particles play a role as a component of interstellar matter, but it has been neglected electrostatic interactions till this day, "although this is difficult to understand, because the electrostatic force between two charged dust particles is almost always greater than the interactive gravitational force" (Goertz, 1991, p. 325).

Transferring the electrostatic interaction energy between dust particles in strongly coupled dust plasmas from small to large scale, there originate an entirely different cosmology. If we regard the *Coulomb's law of electricity*, it comes to light that between two spherical-symmetrically distributed point electric charges, there is an acting force directly proportional to the product of the magnitudes of each of the charges and inversely proportional to the square of the distance between the two charges (spherical epicenters).

The result of the electrically charged bodies in the same direction is then, that a interrelated repulsion of the planets takes place and stable planetary orbits are positioned. But a result is also, that mainly takes place a pressing of materials from outside against the planet. As against, gravitation acts as a pulling force and assume a role which is hardly to measure as against "Coulomb's pressing", because the electric effect is much stronger

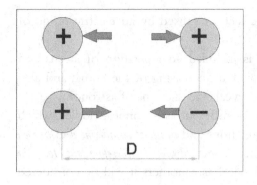

Figure 41a: Coulomb's law. *Diagram describing the basic mechanism of Coulomb's law; like charges repel each other and opposite charges attract each other.*

than the gravitational, although formally the Coulomb law and the gravitational law have the same mathematically structure. In other words, we are pressed and not pulled. Therefore, the "gravity" can be eliminated through appropriate physical "tricks". This view is described as an "electrical gravitation" in my book *Mistake Earth Science* in the chapter *Electrical Solar System* (Zillmer, 2007, p. 234–251).

Since electrical interactions are considerably greater than the mass forces of attraction (gravity), the currently recognized standard model of cosmology must also be wrong, because this requires significantly more mass and energy than actually observable and provable. In other words, it is *far too little* material for supposed gravitational effects existing in the whole universe. Therefore these not yet observable masses and energies typically to be termed "dark masses" and "dark energies" and will be searched feverishly. This search consumes huge amounts of taxpayers' money, including the money-devouring particle accelerators. If this fails, what is to be expect, change of the *Einstein field equations* (EFE) is proposed – which are a set of ten equations in Albert Einstein's general theory of relativity which describe the fundamental interaction of gravitation.

Einstein himself had included the cosmological constant in his original field equations that require a static universe. However, after the Big Bang theory, and thus an instable-expanding universe were introduced in cosmology, Einstein initiated to remove the cosmological constant in turn. Today it is again enthusiastically debated and it circulating several models for the development of the universe according to the Steady State theory, which is a model, developed in 1948 by Fred Hoyle, Thomas Gold, Hermann Bondi and others as an alternative to the Big Bang theory. If one rejects the Big Bang hypothesis, which is supported by the discovery of the cosmic background radiation in 1965, and takes into account interstellar matter in the plasma state by electric effects as discussed already, then one needs not any (fictitious) "dark mass" nor "dark energy". One could then argue that the feverishly searched "dark energy" is the energy contained and transferred in the plasma of the universe. The cosmic background radiation, whose asymmetric distribution runs afoul of the Big Bang theory (see *Sterne und Weltraum*, April 2007), is an electromagnetic radiation, which cannot assign to a specific source.

So you need no "dark energy" and "dark mass", because the plasma-electric effects and forces are much more stronger than gravitational effects. Therefore less energy is required for the steadiness of the universe, because between two electrostatic loads (for example Sun and comets or planets), there is a strong force, called Coulomb force in the principles of electrical engineering.

Such an electrical force in the universe was once seriously discussed, but then denied, and from the cosmology relegated as "Free Energy" to the esoterism. In the meantime, however, the effect of energy in the universe is being discussed which bears different names, but generally it is called as vacuum energy or zero point energy. This cannot, however, be considered in the modern cosmology as per the Kant-Laplace theory, since it is contrary to Einstein's general theory of relativity, because the total energy of the universe is supposed to equal zero. Vacuum energy does not work together with the gravitational force, because it is too weak to constitute as the main effective force in the universe.

An essential point is that vacuum energy can originate real particles of the standard model of elementary physics from nothing – in the otherwise empty room. One could say with Einstein's words that matter arises from energy. This matter is well structured in the plasma state by sufficiently large electrostatic interaction forces. If dust particles of interstellar matter are electro-statically charged and should to be contained in a neutral charged cosmic dust plasma, then regular, crystal structures could be created as studies have shown.

In this way the formation of dust rings around Uranus and Neptune can be explained also (Goertz, 1991, p. 327 f.). These rings are formed by electrostatic interactions in strongly coupled dust plasmas (Ikezi, 1986, compare Schweigert / Schweigert, 1998). And due to the electrostatic nature they are stabilized over long periods, as long as there are no changes to the field strengths. If you want to explain the existence of the dust rings mechanically, taking into account the gravitational effects, you must come to the conclusion that these gracile structures are short-living phenomena.

Cold Comets

Let us now go back again to the electrical anode-cathode system with reference to comets. We searched for hydrocarbons, primarily for methane. If many of these volatile substances are emitted; then a comet must be composed of a *low-temperature condensate* because such volatile materials do not withstand high temperatures. Such a requirement corresponds to an (already mentioned) meteorite group called carbonaceous (carboniferous) chondrites (Wilkening, 1978). These contain up to three and even five percent carbon by weight, which exists in the form of *non-oxidized* hydrocarbons (for example graphite). Of these, only a few percent consist of carbonates in the form of carbon compounds. Moreover there are organic compounds such as amino acids and it is to register a high water content. As such meteorites consist of such many volatile components; they may *never have been heated during its period of existence*. In the ice-cold universe this condition is fulfilled definitely. Comets must therefore have originated from a low-temperature condensate.

But what happens when the cold comet comes to the Sun so closely, that ignition takes place? Doesn't the low-temperature condensate get destroyed by the heat? Let's look comparatively at the glow discharge in the gas-discharge tube. We recognize that the cathode itself remains basically cold and the heat development takes place *before* the cathode at a certain distance from the comet nucleus. The development of heat directly at the core of the comet is thus restricted only to *limited* areas, while the nucleus otherwise remains cold like the cathode in the gas discharge tube (Fig. 37. page 92).

If we are so far consistent with findings of experimental physics, then we now look to the anode, in our case the Sun. Because we are dealing here with a glow discharge also, there is only one conclusion possible: The Sun itself must be as cold as the comets nucleus and just is not extremely hot as one claim to believe – if the comparison with the gas-discharge tube and the anode-cathode-principle is correct.

But the Sun is supposed to be 15.6 million Kelvin hot at its core! How do you know that? Simple answer: Like so many seemingly established facts, this is a pure thought or calculation model. Since one imagines a huge nuclear fusion inside the Sun, in which supposedly 564 million tons of hydrogen are fused into helium per second, the temperatures inside the Sun are calculated arithmetically. In other words: nobody is capable of measuring the temperature, or even looking inside the Sun!

In fact, the calculation of high temperatures is not reached even once through classical calculations, because the resulting kinetic energy of protons turns out to be *too cold* for a nuclear fusion, as pressure and temperature inside the Sun are not sufficient to satisfy the conventional theory. But, for this to take place anyway, at least mentally, one uses a trick called *quantum–mechanical tunnelling effect* – often discussed as *barrier penetration*. This refers to the quantum mechanical phenomenon where a particle tunnels through a barrier that it classically could not surmount because its total mechanical energy is lower than the potential energy of the barrier, just like the so called Coulomb-barrier due to electrostatic interaction that two nuclei need to overcome so they can get close enough to undergo a nuclear reaction. This energy barrier is given by the electrostatic potential energy:

Since we are dealing with special conditions, the probability of a fusion of two hydrogen nuclei in the interior of the Sun is very low, even if we assume that the Sun is actually very hot. However, because an unimaginably large number of nuclei should have been present, one thinks that still huge amounts of energy are released. Through this "limited" nuclear fusion, as researchers admit, the Sun should be using its energy sparingly and over a long period, it must be radiating even a constant amount of energy. In this way a special case raised to a rule and the long-life cycle of the Sun will be also "proven". In contrast to these very special conditions, with the electric model this long-life cycle of the Sun results automatically – like the durability of an anode in a gas-charging tube.

The idea of a hot solar core also results in a direct contradiction. In accordance with the standard model, our present Sun would have increased its solar radius during the period of 4600 Million years nuclear burning by about 12 percent and its luminosity must have decreased by about 28 percent:

"The latter description provides us with still some difficulties in view of geo-scientific knowledge. If one would reduce the present solar luminosity by 28 percent, there would have been corresponding temperature decrease in Earth's surface which would then be covered with a layer of ice. For such a phenomenon, there is no paleo-geological evidence. And vice versa: If the Earth ever possessed a sheet of ice, the high albedo of ice could have also not melted in spite of increasing solar luminosity", William Deinzer, Professor of astrophysics and astronomy at the observatory of the *University of Goettingen*, Germany, give to consider (Deinzer, 1991, p. 5).

The Cold Sun

If an exorbitantly high temperature should prevail inside the Sun – due to a suspected nuclear fusion – the Sun must be hot at its outer edge and therefore in the visible area also. In fact, we see a 400 kilometers thick photosphere around the Sun only .These is not really hot, as one would expect in case of a temperature of 15.6 million Kelvin in the interior of the Sun. In contrast to this suggestion the photosphere is just under 5800 Kelvin (around 5500 degrees Celsius) "cold". This indeed determinable temperature would be only 0.04 percent of the assumed temperature in the interior of the Sun.

As shown in the case of comets, the heat develops outside of the Sun, thus before Sun's surface within the relatively "cold" spherical shell of light, the so-called photosphere. Only further outside in the approximately 2000 kilometres thick layer called chromosphere, which consists of hydrogen and helium, the temperature rises to approximately 10,000 Kelvin. Finally, *solar prominences* rise up through the chromosphere from the photosphere, sometimes reaching altitudes of 150,000 kilometers. These gigantic plumes of gas are the most spectacular of solar phenomena, aside from the less frequent solar flares. Above the chromosphere there is a so-called transition region, where the temperature increases rapidly to the hot corona, which forms the outermost part of the atmosphere. There prevail temperatures of around one million Kelvin. Thus the un is really hot definitely above its surface in the "ball-shaped layer perceived to emit light".

Let us now compare the anode in the gas-discharge tube, the *anode glow* then appears at

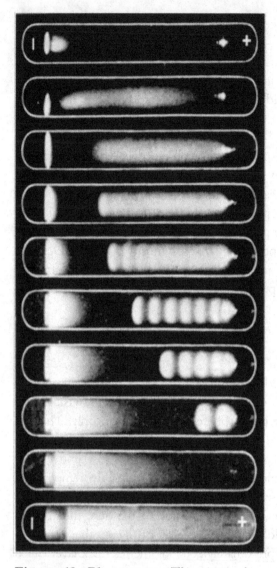

a certain distance from the anode surface (Fig. 39, p. 94). This layer represents

the photosphere of the Sun. Outside, the anode glow is connected with the so-called *positive column* which constitutes the bulk of the discharge. In contrast to a fluorescent lamp the distances in our planetary system are very long in contrast to the size of electrodes (comets and Sun) and there is formed only one positive column outreaching into the interplanetary space. This goes quickly into the cold *Faraday space* which extends to the *negative glow*, the brightest layer before the comet nucleus (= cathode). The sun-comet system corresponds not only to the visual appearance in the gas-discharge tube, but also qualitatively to the temperature profile.

Now we take a closer look at the anode (= the Sun) in the gas-discharge tube. We discover a darkroom, a dark layer between the anode surface (= solar interior) and the bright anode glow over it (= photosphere), the so-called *anode dark space*. Where is the dark, more precisely selected as *less bright* space in case of the sun? Similar to the *anode dark space* in the gas-discharge tube, it must be the layer under the photosphere. This must be the interior of the Sun under the visible bright surface. This layer is dark, but appears pitch black due to the contrast in relation to the brightness of the glow (Fig. 39). On the "spherical ball of light" or the visible solar surface (= photosphere) appear dark spots; their

Figure 42: Phenomena. This picture shows the luminous phenomena in an electrical gas–discharge tube between the cathode (= comet, on the left hand side) and anode (= sun, on the right hand side) in case of decreasing gas pressure. In the planetary system the cathode rays which are directed towards the sun (anode) turn in front of the comet nucleus, surround the nucleus and in this way are able to form a comet tail. Kaufmann: "Handbuch der Experimental–Physik", 1929.

number varies periodically. For these sunspots, there is no satisfactory conventional theory, which explains the appearance and disappearance sufficient. It is believed that sunspots are caused by local disturbances in the huge solar magnetic field. By this are disabled the movements of the "convection cells" suspected inside the Sun, which swirl the heat of the solar interior to the surface. This assumption, however, creates a contradiction, because the temperatures in the range of the black spots are smaller than in the photosphere that surrounds these areas.

What is the solution for the case of the *cold* sun, taking into account the electrostatic anode-cathode model? It is for this case obvious that the sunspots are no spots, but holes. Through these we can see in the underlying solar interior. Immediately below the photosphere is thus a dark room in the form of a spherical shell surrounding the solar interior, which corresponds to the *anode dark space* in the low-pressure gas–discharge tube. Below the *dark space layer* there should be situated the core of the sun, which is cold like the anode in the gas-discharge tube, but this will not to be discuss here in detail.

In contradiction to the conventional model, the mysterious, in the area of the photosphere openings (= black spots respectively sunspots) stronger appearing magnetic fields, are easy to explain, because these are disposed perpendicular to the electric fields and can emerge through the photosphere openings easier. Therefore the magnet fields appear with greater intensity in the range of sunspots.

Since the solid surface of the cathode and anode in the low pressure gas-discharge tube is "cold", the Sun must also be cold below the photosphere or more exactly below the *anode dark space*. The electron and gas temperature are hardly linked and there is no thermal balance. Therewith the Sun is able to build a stable system; this must be in static equilibrium. Positive and negative forces must remain therefore statically balanced. Otherwise the Sun would explode, as is already observed in case of Nova Pictoris or the Nova Hercules.

This means that the Sun is not representing through and through a positive pole (anode). If we consider a static equilibrium of forces, there must be a negative pole (cathode) in the centre of the Sun, according to the electric model. The all-round pressure- and force balance must coercive result in a ball shape at the time of the formation of the Sun, as it is the sole possibility to get a stable static form of all acting forces. The today seemingly only 400 kilometres thick photosphere appears as a *ball shaped layer of light* and is therefore a static equilibrium structure, called by me as dynamic "neutral spherical shell". Within this quasi-static neutral layer zone, the forces attacking from inside and outside are balanced statically. In other words, the kinetic (positive) electricity flowing toward the Sun is in static equilibrium with the potential (negative) electricity, which is located in the interior of the Sun.

Due to the turbulent dynamics within the photosphere or the development of a kinetic instability, the energy in the electrostatic waves exponentially increases until

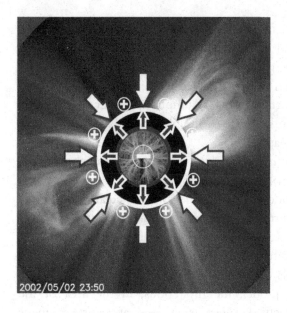

Figure 43: State of equilibrium. The Sun and the planets must be in static equilibrium. The outer and internal (electric) forces are in balance within the "neutral spherical shell" – this is the photosphere in case of the Sun. This may explain why the temperatures in the more external lying corona are about 165 till 330 times higher than in the optical surface of the Sun (photosphere = neutral spherical shell). Without this explanation there is a so-called coronal heating problem caused with the standard model, because the Sun cannot direct heat conductions from the too cold photosphere. Picture background: out flowing solar plasma (SOHO Consortium, ESA, NASA).

saturation occurs. At the same time, the ionized gas respectively the plasma is heated all around the spherical shell and the photosphere begins to shine. There are triggering multiple dynamic processes and events, including so-called shock waves, which are flashing in the interplanetary space.

One shock wave follows the other, and then they fuse these with each other, and in this way a "shock shell" arises around the Sun. This is filled with turbulent plasma. Finally, it creates a shield (= spherical shell) against the external pressure of energy-rich particles of the galactic cosmic radiation (= GCR). "This clarifies the sudden drop in the intensity of the GCR with increasing solar activity as well as the general course of counter-cyclical process of the GCR in the solar cycle. Because, the frequency of such occurrences follows the variances of solar activity very closely" (Schwenn, 1991, p. 38).

We hereby obtain an explanation for the really obvious fact that the Sun does not heat planetary celestial bodies through thermal radiation, but that this activity and interactions are determined in the non-visible energy spectrum. Therefore, in the German book *Plasmaphysik im Sonnensystem* (plasma physics in the solar system) a warning notice is given: "The relative number of sunspots which is usually used as a measure for the solar activity is of a very problematic size. This indicator not even describe the real activity, but rather it is at best, a visual indication of possible activity" (ibid., p. 38).

Climatologists say, however, that the Sun has no or very less influence on the Earth's climate, as, for example, in years with a reduced number of sunspots; the solar radiation reduces at the most only by about 0.1 percent. Such small differences in radiation, say climatologists, is not sufficient to ensure the control of Earth's climate. Absolutely right! But the visible Sun's rays are effectively the only visible expression of largely invisible energy spectrum steered by the Sun. With other words, we can only see a little part of all effective radiated energy. The number of sunspots should be monitored regularly for over 350 years, and a connection with air temperatures measured above the Earth ground clearly identified: In the case of few sunspots it is colder, but with their cluster, it is warmer. The anode-cathode model can also explain why the number of sunspots follows a cycle. If there are no holes (= sunspots) in the photosphere, this *neutral spherical shell* is enclosed and the activity of the Sun is low: Because less energy can get out of the Sun interior. But if there are numerous holes, then the Sun is active because in order to maintain the static equilibrium and the stability, the openings must be close. If this fails, the star will burst.

If there are appear and increase many sunspots and thus holes in the photosphere, radiation and electricity escape from the solar interior into the planetary space. Thus, electrically conductive celestial bodies will be supplied with more (emerging) energy, and it is getting warmer, for example on Earth. In the opposite case with few sunspots, the photosphere is more closed, and so the energy is more trapped inside the Sun and it gets colder on Earth and other electrically conductive celestial bodies.

The already mentioned, the climate Echo G-model was first presented in Hamburg in 2003. It shows that the Sun has had a major influence on the climatic conditions on the Erath and volcanic eruptions contribute to short periods of cooler temperatures. "This modelling coupled with an atmosphere-ocean model fits very well with the large number of individual reconstructions of the past 1000 years" (Berner / Streif, 2004, p. 220 f.). The Sun affects the Earth and other planetary celestial bodies with the electrical energy transmitted through the plasma. At the same time, the universe is bitterly cold and the Sun doesn't warm the interplanetary space, instead it will only warm the spaces on and before the electrically conductive celestial bodies – electrostatically like we can observe in a gas-discharge tube.

With the help of the electrical model, one could explain why comets suddenly shine bright and at the same time the space between the Sun and planetary celestial bodies is almost definitely bitter cold and very dark like in the gas-discharge tube. Electrically non-conductive celestial objects therefore appear dark and can only be detected if they move in front of bright objects. But the Moon is *not* electrically conductive and yet appears to be bright at certain times. Why this can happen, if and when the planetary space between the Sun and the Moon is dark?

As recently as 2007 it was clarified that the Moon's surface is electrically charged by analysing data of the probe *Lunar Prospector*. The magnetic tail (magnetotail) of the Earth flutters in the direction of the solar wind. When the Moon passes through this terrestrial magnetotail, it comes in contact with a huge plasma layer of charged particles, making the surface of the moon negatively charged and electrostatically charged dust particles are flowing over the lunar surface. There is even a whole kind of dust storm towards the stronger charged night side. The Moon comes into the magnetotail three days before becoming fully bright and it takes about six days to cross it, and to pass out on the other side, explained the researchers (Halekas, 2007). This means nothing else than that the Moon is glowing when our satellite is located right in the magneto-tail, and becomes dark, when the satellite comes out of this. Also, there is a type of dust storm towards a strongly loaded night side. "The Moon enters the magneto tail three days before that is bright all in all and it lasts around six days for it to cross and to come out on the other side" the researchers informed. (Halekas, 2007) This means nothing else than that the Moon shines bright, when our satellite is positioned in the magnetotail, and dark, when it emerges from this.

Without magnetotail of the Earth, our Moon would be dark, because contrary to the Earth, it does not have any strong magnetic field. The fully ionized solar wind plasma does not recognise the case of an electrically neutral Earth moon as an obstacle at all. Therefore, in contrast to a conductive celestial body the plasma collides without forming a bow shock and there is no appearance of light – the moon is dark. If the moon's surface will be electrically charged while moving through the terrestrial magnetotail, because it will be generates an electric potential transversely to the magnetotail of the Earth, a glow discharge is formed as soon as the electricity increases like an avalanche and the ignition voltage is reached.

Neither the Sun nor other stars illuminate or warm the dark and ice cold universe directly because they are only glow discharges. Electromagnetic energy is radiated in the *invisible* spectrum and electricity over the plasma. Nevertheless, there are phenomena of light, for example, in the form of luminous fogs which owe their brightness only to the features of the plasma, not because stars light up the universe. The universe is pitch-black as we assess easily while looking in the night sky! Here, we are talking about electrical or electromagnetic phenomena, for example "plasma clouds" which light up in our solar system or in the rest of the universe.

Theoretically, the Sun cannot illuminate interplanetary space directly because the universe shall be a vacuum so there is no medium which could transfer the Sun's rays. Therefore, it was previously considered that a very delicate substance named "ether" must have existed in the universe. For our considerations, we do not need a finely dispersed medium (= ether) in the universe, how may has been clarified only now. This "ether" would, after all, slow down the motion (translation and rotation) of all bodies, because these would move within something: As a result a moment of inertia is created from the mass, whereby the movement of celestial bodies would steadily slow down.

The transmission of electromagnetic energy and electricity takes place over the unevenly distributed plasma. If one observes in the universe for example a black area

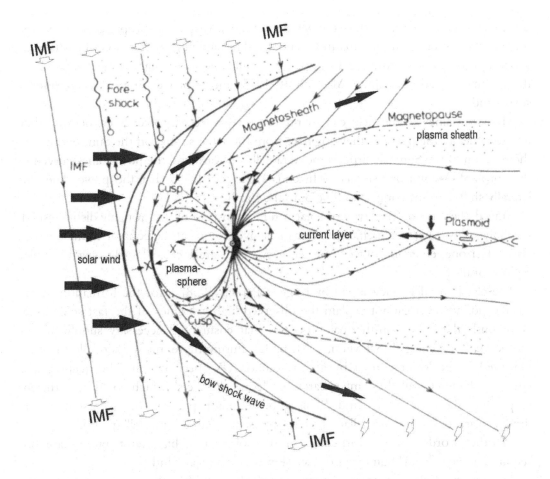

Figure 44: Side view of the magnetosphere. *You can see the bow shock, the magnetopause as the abrupt boundary between the magnetic field and surrounding plasma. The magnetopause ripples, flaps, and moves inward and outward in response to varying solar wind conditions. If the interplanetary magnetic field (IMF) is directed away from the Sun, "the field lines coming from the Ssun should be connected to the Earth's field in the northern polar cap, while the south polar cap is connected with the magnetic field lines going into the interplanetary space. Only in the northern polar cap, in this case, cosmic rays coming from the sun should penetrate" (Scholer, 1991, p. 128).*

without shining stars, this is simply an area without any plasma. Consequently, through this absolutely empty space, no starlight's respectively energies and currents can be transmitted. In a region without plasma, it is simply black. Instead of a universe filled with ether, let us consider the plasma state and thus an energy level of the universe.

Our entire solar system is filled with solar wind plasma. A variety of plasma physical processes are occurring, such as the generation, attenuation and propagation of waves, wave-particle interactions, plasma turbulences and instabilities, acceleration of particles or shock-free waves (see Schwenn, 1991, p. 20). Finally, the solar magnetic field is frozen in the plasma (see Kippenhahn / Möllenhoff, 1975) and is pulled outward by expanding solar wind.

The streamlines of the field can therefore be regarded as power lines, because they link together all the particles that originate from the same source on the Sun. Because of the rotation of the Sun, all particles move radially outward and the power lines curved in the form of Archimedes spirals (Schwenn, 1991, p. 19), so that the power lines hit laterally shifted to the connection line onto the Earth.

Indeed, there is no single type of solar wind, but instead of this, roughly distinguished, not only a slow one, which emerges over the active regions at the equator of the Sun but also a fast one, emerging out of corona *holes*, which are located largely over the polar regions of the Sun.

The classic model – developed by Eugene Parker (1958) – for the formation of the solar wind, however cannot explain the observed high velocities of the fast solar wind. "Obviously there are non-thermal processes that ensure the necessary supply of this momentum and energy" (Schwenn, 1991, p. 42). Furthermore, many *jets* of plasma were observed in the lower corona by high-resolution UV-spectrograph. The quantity and speed of these jets called as "microflares" or "coronal bullets" which could be sufficient to produce the entire solar wind. "This quite serious hypothesis of not just thermally driven solar wind is of course diametrically opposed to Parker's model" (p. 45).

In other words, for the formation of the solar wind, high temperatures are not required. Even a "cold" Sun may produce slow and fast solar wind.

The stream of charged particles coming from the Sun hit with supersonic speed onto the planetary obstacles. A bow shock is created like an aircraft flying at a supersonic speed, if the planet possesses a magnetic field and an electrically conductive atmosphere. The flow energy of the cold solar wind plasma is not only changed into thermal energy at the bow shock wave, which is formed in front of the Earth (Figure 44), but also, the cold stream of ions and electrons are turned vertically to these in an opposite direction as they enter the Earth's magnetic field. "Thus, flowing electricity is created in the boundary layer" (Scholer, 1991, p. 122).

Because the solar wind distorts the Earth's magnetic field, with this are associated electric currents on the dayside according to the Maxwell's laws. Therefore, the night–side distortion of the geomagnetic field to a tail-like magnetosphere is accompanied by electric currents, that power can be up to several million amperes (ibid., p. 106). Another important power system is the ring current respectively electric current which consists of a band, at a distance of 2 to 9 Earth radiuses, which lies the equatorial plane. The ring current energy is mainly carried around by the ions, most of which are protons flowing

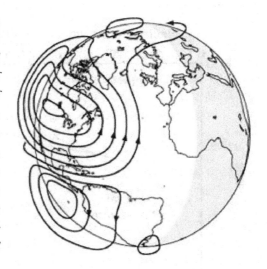

Figure 45: Ring currents. Electric ring currents produced by the Sun (equatorial electro–jet) on the dayside of the ionosphere (Figure: United States Geological Survey). The electrostatic field of the Earth results from an electrical excess charge of the Earth's surface caused by ionizing radiation from space. Since the electric potential on the surface of a conductive body is distributed evenly, the conductive human body feels no voltage difference in the air, but sometimes, a small force between bodies with different charges. There can be an indirect influence on the human body, since the air ionization decreases the physical and mental performance.

along the Van Allen belt in east-west direction (Baumjohann, 1991, p. 106 f.). "We may not have the complete understanding of our own Earth…" (ibid., p. 190).

The Sun produces different electrical systems on Earth, which are not limited just to the magnetosphere. There are other power systems, which currents flow not only vertically to the magnetic field lines, but also parallel to them, often called as Birkeland currents named after Kristian Birkeland (1867-1917). These are connecting currents in the magnetosphere and its boundary layers with those in the polar ionosphere and thus allow an exchange of energy between these regions. In almost all cases, the generator is located in the boundary layers of the atmosphere to source its energy from the kinetic energy of the oncoming solar wind. overall, a technically separately exited dynamo phenomenon occurs:

"The turning moment to move the inductor in a dynamo is delivered from the solar energy. The magnetic field corresponds to the geomagnetic field of the Earth. The load resistance of the dynamo corresponds to the reciprocal electrical conductivity in the ionosphere. The self-induction of the dynamo inductor corresponds to the charge separation and thereby produced polarization field. The inductor, which is moved, represents the wind and the inductor power represents the electric current in the ionosphere. Hence the names dynamo theory and dynamo region (Volland, 1991, p. 294). In contrast to this explanation, conventional thinking geophysics are searching an independently-starting and self-sustaining dynamo effect in the Earth's core.

It is understandable that the electric current driven by the convection in the dynamo region of the ionosphere produces (joule) heat, which heats the entire thermosphere

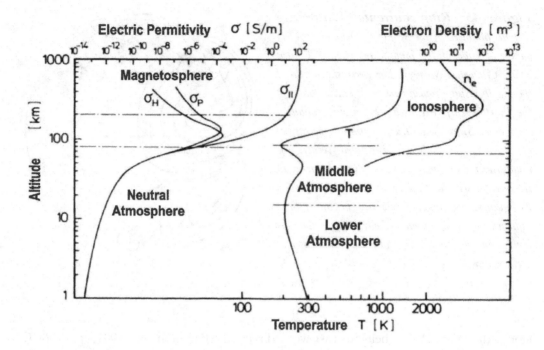

Figure 46: Average altitude profile of the electrical conductivity, temperature (in Kelvin) and the electron density. *Here some typical numerical values given for current, voltage and power of energy sources in the atmosphere, which can form large-scale electrical (potential–)power fields and currents. After Volland, 1991, p. 285.*

located at a height of 85 to 600 kilometers. This source of heat supplies a certain part of the warmth caused in the same area which is heated due to the solar energy and can exceed these even during disturbed geomagnetic conditions. As a result, the temperature in the upper border of the thermosphere (towards the exosphere) which varies typically from 700 to 1500 Kelvin depending upon the solar activity; can then quickly increase to a temperature of several thousand Kelvin.

The magnetosphere protects the Earth against the charged particle stream coming from the Sun. Even though there is no hard border, but in the underlying ionosphere occur the ionization of gas molecules. By ionization one understands the process in which at least one electron is removed from a neutral atom or molecule. It can be stated that the ionosphere exists only due to the solar radiation activity, because the ionization takes place above an altitude of about 80 kilometres mainly by solar respectively extreme ultraviolet and X-ray radiation. For this reason the energy bursts of the Sun affect the active state of the Earth directly.

Therefore, in contrast, the conventional model based on a structurally homogeneous ionosphere is not matching the reality. This is reflected in the deviations from the generally expected behavior. Thus, the maximum of the electron density is located in the early afternoon and is not consistent with the time of highest position of the Sun (day anomaly). Despite the lack of sunlight, the ionization increases during the night (night anomaly). In the absence of sunlight, there are strong ionized layers over the poles during the polar night (polar anomaly). In winter, the electron density is higher than in summer (seasonal anomaly). The maximum of the electron density is not above the equator (equatorial or geomagnetic anomaly). In addition to these consistently observed abnormalities, there are also spontaneously occurring, short–term disruptions in the ionosphere, most of which occur due to the solar burst and uncommon radiations.

The universe is ice cold! A warming take place around the Earth and is mainly caused by the solar energy supply (*not direct* thermal radiation) as well as cosmic radiation and the result is a formation of electrical systems near the Earth. Accordingly, a warming occurs on the dayside, but not because of the heating up of the Earth by the Sun, but because here the not visible *energy* emitted from the Sun enters the Earth's magnetic field. This leads to a pressure difference between day- and night side, and it inevitably creates a permanent wind system between the two hemispheres, which will balance out the prevailing temperature and pressure variations.

In this way, we have described the source of a large-scale quasi-static electric field in the ionosphere. The tidal winds and magnetosphere storms diversify the terrestrial magnetic field which is subject to the interaction of the solar wind with the magnetosphere and its strength is therefore influenced by the Sun. The same procedure takes place in case of the Sun, as it also forms a bow shock wave namely in the intergalactic plasma and therefore in the galactic current of charged particles. The Sun processes "galactic cosmic energy", which is modified and emitted into the planetary system.

Our planetary system and thus our Earth is therefore subject to (superordinate by) a "control mode" due to the galactic cosmic radiation, thus the state of stress and strength of energy can change and for which reason, for example, the distances between the planets and the Sun and therefore between the planets to be subject to change. The planetary orbits can therefore vary as described by our ancestors in the traditions.

Electrical Thunderstorms

Since the air is a poor conductor of electricity near the Earth, electric fields and their effects on the atmosphere and thus on our own lives have been ignored long time or marginalized as esoteric ideas. The electrical conductivity decreases a lot below the ionosphere, but this is sufficient near the Earth to let an electric voltage applied to the plates of an electrometer-scale collapse within a few minutes. Therefore one should not neglect the plasma components of the air in any way, as for example the modern climate models doing.

The "electrical conductivity of the air is the requirement to develop large-scale electric (potential) fields and currents in the atmosphere. For the emergence of such fields however, non-electrical forces are necessary. Currently, we know three processes that produce these large-scale electric fields" (Volland, 1991, p. 285):

- Tidal-like winds in the dynamo region (ionosphere)
- Interaction between solar wind and magnetosphere plasma
- Thunderstorm activity in the troposphere.

Figure 47: Electrical current. *The electrical current in the ionosphere especially in the dynamo region has a dependency on the solar activity through which the weather (including thunderstorms, hurricanes) is affected.*

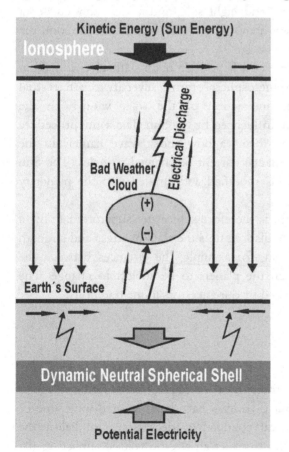

Overall, "there is some similarity with a technically externally caused dynamo. The turning moment to move the inductor in a dynamo is supplied by solar energy ... The moving coil represents the wind and the inductor current represents the electric power in the ionosphere" (Volland, 1991, P. 294).

Between the surface of the Earth and the highly conducting ionosphere, there is very strong electrical voltage drop, by which Earth's surface and upper atmosphere are comparable to a giant spherical capacitor in their function (Haber, 1970, p. 82).

In fair weather regions, ionosphere electrical power flows towards the relatively well electrically conductive surface or dynamic neutral spherical shell.

The discharged electrical power flows from the electrical capacitors in the crust, sometimes triggering earthquakes, during thunderstorm activity upward to the cloud, and from there up into the ionosphere, thereby sometimes causing strange luminous phenomena, sometimes mistakenly described as huge UFOs.

The first two processes which run mostly in the upper atmosphere respectively ionosphere and magnetosphere are already described precisely in the earlier chapters. What is the case with the thunderstorms?

The electrical conductivity of the air which increases almost exponentially with altitude and is influenced by the Earth's magnetic field at an altitude of above 80 kilometers is a requirement to develop large-scale electric (potential) fields and currents in the atmosphere (Fig. 46, p. 110). These are generated above in the magnetosphere the interaction between solar wind and magnetosphere plasma. In the underlying dynamo region belonging to the ionosphere provide tidal-like winds for the maintenance and development of such electrical effects. In the lowest layer of the atmosphere, (troposphere) thunderstorms are responsible for an atmospheric electricity current flow. (Volland, 1991, p. 284).

As the air near the Earth is a poor conductor of electricity, we excluded the possibility of long–time electrical effects in the lower atmosphere completely. However, thunderstorms are electrical, strictly speaking, electro-magnetic phenomena. They are characterized by the fact that their source strength is sufficient to activate electric gas discharges. About 2000 thunderstorms happen on Earth at the same time, covering an area of about 0.1 percent of the Earth's surface, whereby each thunderstorm discharges electrical power to its environment (ibid., p. 290).

Between the surface and the highly conducting ionosphere, there is a very strong electrical voltage drop, by which Earth's surface and upper atmosphere are comparable to a giant spherical capacitor (Haber, 1970, p. 82). The electric potential between ionosphere and Earth's surface can fluctuate to 100 percent or more (for example see Mühleisen, 1977; Markson / Muir, 1980).

Earth and ionosphere behave like electrically equipotential surfaces from which electrical power flows through areas of relatively good weather to the electrically conductive crust. A voltage balance occurs by a discharge current which flows from the electrical capacitors in the crust towards the bad weather cloud during a thunderstorm activity and from there up into the ionosphere (see Fig. 47), whereas the conductivity of the air increases almost exponentially with the height and the path of least resistor is sought. These are sometimes strange luminous phenomena caused between clouds and ionosphere.

Sprites and Elves

Since the current in a lightning discharge usually flows from the Earth's crust to the cloud, there are corresponding effects *above* the clouds up to the ionosphere, but these have been targeted photographed only since 1989 from airplanes and space shuttles.

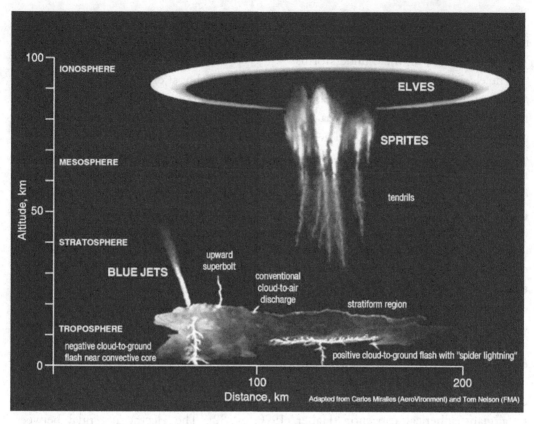

Figure 47a: Transient Luminous Events. *Above large thunderstorms occur other kinds of electrical phenomena called transient luminous events (TLE's) like red sprites, blue jets, and elves.*

Thunderstorms above the clouds were in the spotlight in 1987, when a rocket from NASA triggered a lightning while reaching higher air layers. The electronics were destroyed by the electric discharge and one had to detonate the rocket. It is now known that rockets create a line style ionized air channel, going behind several miles, in which long line-type lightning's can emerge. Also at the entrance of meteors in the atmosphere it has been observed that flashes follow in the meteor trajectory. These lightning's caused by external events correspond to lightning that are produced by a thunderstorm above the clouds. At an altitude of about 70 kilometers *upper-atmospheric lightning* or *upper-atmospheric discharges* are phenomena that occur well above the altitudes of normal lightning. The preferred current usage is *transient luminous events* (TLEs) to refer to the various types of electrical-discharge phenomena in the upper atmosphere. These phenomena are well known in the aviation; however, they have been dismissed as imaginations of the pilots because they could not be explained in a conventional manner.

114

Some symptoms were described as UFOs, and many sightings were not revealed, until only in recent years, when the electrical nature of such phenomena was recognized.

So-called sprites often appear reddish-orange or greenish-blue in colour, are large-scale mostly column-forming, but may look like an atomic cloud. Rarely do they appear in the form of board fences. They often occur in clusters, lying 80 to 145 kilometers above the Earth's surface. Sprites have been held responsible for otherwise unexplained accidents involving high altitude vehicular operations above thunderstorms.

But there are other lightning phenomena in even greater heights. In addition to sprites there are so-called elves in the ionosphere. These lightning discharges occur in about 90 to 100 kilometers above the great storm clouds on the edge of the ionosphere as a reddish ring. Elves often appear as a dim, flattened, expanding glow around 400 kilometers in diameter that lasts for, typically, just one millisecond. Their colour was a puzzle, but is now believed to be a red hue. Elves were first recorded on a shuttle mission on October 7, 1990 in an area off French Guiana.

One will be able to classify these electrical phenomena only when the Earth will considered as a spherical capacitor according to Heinz Haber (1970, p. 82), and lightning phenomena and also hurricanes should not be viewed as random events or as caused by global warming.

Let us now consider the normal current flow *below* the clouds. The current flow is usually directed upward, therefore from the Earth to the negatively charged lower edge of the cloud. But running thunderbolts rather than vice versa, from the cloud to the Earth's surface?

If local field strengths are formed within a thunderstorm, an electron avalanche can arise, which forms a conductive channel. This proceeds down in the form of steps and reaches the surface of the Earth or outstanding objects on the Earth's surface. This discharge is known as the cascading partial discharge. After the contact, the electrons stored in the channel flow to the ground thereby forming an electric current directed to the cloud. It heats the plasma in the channel to 30 000 Kelvin, and there occur ionization and recombination. We see the recombination glow as lightning discharge. This main discharge is the negative lightning, also known as cloud-to-ground lightning. Pre– and main discharge can repeat several times, so a flickering lightning is seen. The increase in pressure in the channel ultimately results in an acoustic shock wave that we hear as thunder (see Volland hear, 1991).

During the pre- and mid discharges, electromagnetic waves are created. The interesting thing here is the antenna theory which has been researched at the University of Florida (Moini et al., p. 1997). It models the lightning channel as a lossy, straight monopole antenna standing vertically on an electrically highly conductive surface. This antenna is fed by an external power source (electromagnetic wave) over the atmosphere and lightning discharges occur (ibid., p. 149, and Moini et al., 2000, p. 29 693).

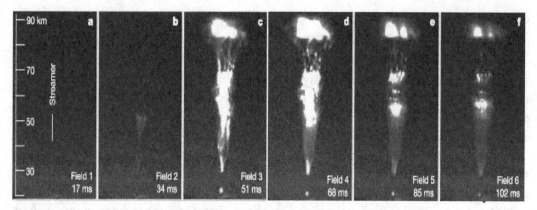

Figure 48: Sprites. *From the top of a huge thunderstorm clouds, a column-like lightning can occur as electrical discharge called sprite, which regulates a difference of up to 300 000 volts between the surface of the Earth and the upper atmosphere (Nature News, 18 Oct. 2001).*

Is it possible that a metallic golf club act as such an antenna, which scatters electromagnetic fields? In Colorado, there are annually several lightning deaths under a *cloudless* respectively blue sky, and in Florida golfers have been repeatedly violated from lightning and sometimes killed, though no cloud was to see in the blue sky. Such local electric discharges are currently ignored.

Despite many new insights, lightning's are due to electro-magnetic discharge processes, which represent an unexplained natural phenomenon. Only recently it has been recognized that the light of lightning is caused through the formation and heating of plasma. The plasma component of the atmosphere is based on the electrical conductivity of the air, by which some phenomena can be explained, which constitutes an absolutely unsolvable riddle in the event of an (absolute) electrically neutral atmosphere.

Nevertheless, until now, it is completely unclear why with a lightning arrives more energy on the Earth's surface than was contained in the cloud before. The explanation offered today is a so-called *runaway breakdown*. According to this, the electron, after reaching a certain critical speed, becomes faster, but this is not sufficient for the strength of the measured radiation. Is it perhaps differently, to see just the opposite, and the cloud arrives less energy because these is emit in the ground below?

In 2004, researchers at the University of Florida in Gainesville have demonstrated that with the cascading partial discharge, strong X-rays are emitted, but apparently not with the subsequent main discharge. The emergence of a flash is *no simple discharge* or a spark discharge, like previously believed.

As it became clear, the lightning does not depend only on the conditions in the atmosphere (thunderclouds), but also on the circumstances respectively the electrical

116

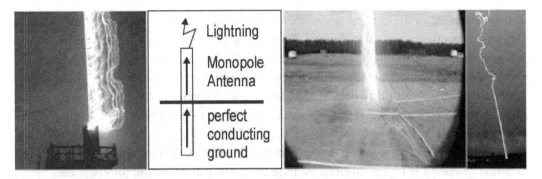

Figure 49: Discharge. *At the University of Florida, lightning phenomena will be experimentally produced, which are ignited by a monopole antenna, which is arranged on a well conducting substrate.*

conductivity in the substratum, and on a larger scale also depend on whether it is an ocean or a continent. On a smaller scale, there are areas in which the statistical frequency of lightning is two to three times higher than in surrounding areas. For example, in southern Germany it may be located less than one flash per square kilometer, in Austria and northern Italy one to two and in Slovenia about three. In addition, there is more lightning in cities and the lightning frequency also depends upon the season. In addition, the geographical latitude plays a role, because the so-called "keraunic level" indicates the number of thunderstorm days per year in a given area. This is between 1 in Antarctica and the Arctic Zone and up to 180 in the vicinity of the equator.

Are there perhaps even preferred thunderstorm routes? If we plot the number of thunderstorms per year on a map, according to the keraunic level, then there are areas with a constant frequency of storms. Combining this, then there are lines with the same flash frequency, the (in Germany) so-called isokeraunen, comparable with the isobars (lines of equal air pressure on a weather map).

Certain localized points on the Earth seem to be particularly good for a current flow from Earth to the cloud due to their conductivity. During this, the negative pole (thundercloud) moving over a conductive ground closes the contact with the positive pole on the surface or in the substratum, consequently lightening is starting.

This brings us to the riddle of the formation of hurricanes. Is it true that actually only about 28 degrees warm ocean water is necessary, or, since this explanation is completely inadequate, are fictitious seeming *wind shears* the activator – as many climate specialists nebulous speculate? In 2005, in Costa Rica, the weather aircraft ER-2 of the NASA measured extremely strong electric fields of the severe hurricane Emily. One can describe a hurricane also as a potential vortex which produces a standing electric field perpendicular to it. In this way, a conducting air channel is formed. The violent lightning in a hurricane represents a puzzle for conventional research. But here we can explain that this concerns natural electrical phenomena which are originally caused by the Sun or triggered from the cosmic rays (Pérez - Peraza et. al., 2008). A 2010-report correlates low sunspot activity with high cyclonic activity. Fewer sunspots appear to decrease temperature in the upper atmosphere, creating unstable conditions that help create cyclones. Analysing historical data, there had been a 25% chance of at least one hurricane striking the continental US during a peak sunspot year; a 64% chance during a low sunspot year (*Florida Today*, 1 June 2010, p. 1A f.). These correspondents with the explanation in this book given before, that the activity of the Sun respectively the radiating of energy is higher, if there are more holes (sunspots) in the photosphere!

We our self can influence neither frequency nor the intensity of hurricanes in a measurable scale. In this regard let us consider another unsolved lightning phenomenon.

Phenomenon ball-lightning

The up to 50 centimeters large ball- lightning's are rare phenomena. Known witnesses were the philosopher Seneca, Henry II of England or physics Nobel prize laureate Niels Bohr, which they described as floating, luminous balls of orange to yellow, reddish and bluish white colour.

Sometimes a ball-lightning sprayed sparks, accompanied by crackling noises. The luminous phenomenon can persist from several seconds to some minutes. It has been described as moving up and down, hovering and moving with or against the wind; unaffected by, or repelled from buildings, people and other objects. Some accounts describe it as moving through solid masses of wood or metal, and this is usually done without leaving damage. Sometimes there are burn marks on pieces of wood for example. In some cases, a ball-lightning was sighted in a submarine boat (Silberg, 1962) or penetrated into an aircraft cockpit and was leaving at the rear end (Singer, 1971). Most ball-lightning's disappears after a few seconds, explode in rare cases, or disappear simply.

There are now various theories that are based either on an inner energy (burning, plasma, vortices, charge separation, nuclear power) or on an external energy source (magnetic or electro-magnetic fields, Earth's magnetic field, cosmic rays, antimatter) (Rakov / Uman, 2003).

Ball lightning's have a clear-cut, closed structure, which, according to eyewitness reports, sometimes is surrounded by a distinct corona discharge, a phenomenon reminiscent of the outer structure of the Sun. Like these, a ball lightning must carry the energy source inside itself. Because these seem to be an electrical or electromagnetic phenomenon, the glowing sheath in turn could represent a dynamic *neutral spherical shell* between kinetic and static energy (electricity). Therefore, extremely high temperatures are *not* necessary to produce the luminous shell. Therefore a ball-lightning as cold as the Sun?

A notable example is found in plasma physics: When a gas is supplied with a sufficiently large electrical pulse (energy), then plasma is formed. If the energy increased further, then the electric charges are subject to a turbulent motion, then sometimes a metastable, vortex-shaped ring is formed. This coherent structure of plasma and magnetic field is known as plasmoid (Bostick, 1957, and Wells, 1970). This, according to the theory of Antonio F. Ranada (2000) should be consisting of closed electric and magnetic field lines, which means that the energy is materialized, because closed field lines produce mass (see Meyl, 1999, p. 6 ff.) The interaction of cosmic electricity and matter opens up the opportunity to develop through the production of ball-lightning in the lab completely new energy sources.

Already Nikola Tesla (1956) produced ball-lightning by large flat coils (Tesla coils) in the 19th Century in a steamy vortex spherical. Plasma balls similar to the ball-lightning discharges have been created by German scientists. Professor Dr. Gerd Fußmann (Humboldt University Berlin) explained in a press release dated 8 May 2006: "Why actually the luminous phenomena occur, is still far from clear. They are visible for only about 300 milliseconds after the current is stopped and the energy supply is thus cut off. Actually, they already should be extinguished after a few milliseconds. Furthermore, the plasma glows very brightly, although the plasmoids appear to be rather cold: A sheet of paper placed above them does get lifted, but doesn't burn." It is interesting that these plasmoids, like the ball-lightning, are cold.

Ball- lightning's can be an example that energy and electricity can materialize and thus occur to be visible. Is it possible that cosmic energy materialize in the interior of the Earth in a similar way? Before we pursue this matter, let us take a look at the current flow in the Earth.

Current Flow in the Earth

The torque for moving the conductor in the ionosphere dynamo of the Earth is supplied by solar energy. The magnetic field corresponds to the Earth's magnetic field and this has a regular variation on the surface of the Earth with the period of a solar day (Volland, 1991, p. 294). In fair weather regions current flows in the direction of the Earth's surface (ibid., p. 290) and into the crust which is relatively a good conductor. In the Earth's crust electric fields are present:

"The geophysicists discovered layers inside the Earth with increased electrical conductivity. It increases up to a depth of 100 kilometers. Beyond, with increasing depth, it then decreases again. In connection with the mixed layers of different electrical conductivity, Pospelow assumes that the shells of the Earth's crust and the mantle of the Earth are electric capacitors. Their (parallel–)plates are different layers of rock.

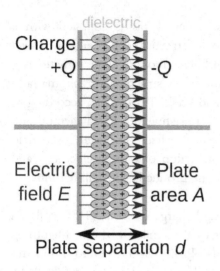

Charge +Q   -Q

Electric field E   Plate area A

Plate separation d

Figure 49a: Capacitor. *Schematic of a parallel plate capacitor with a dielectric spacer. Two plates with area A are separated by a distance d filled with dielectric material between. When a charge +/–Q is moved between the plates, an electric field E exists in the region between the plates. The dielectric material becomes polarized due to the charge displacement, reducing the total internal field, and increasing the capacitance. Autor: Papa November.*

In these, an accumulation of electric charges takes place and from time to time they are pierced by underground-lightning's (...) geological resting occurs if an accumulation of electricity takes place, whereas active periods correspond with the discharge of the capacitors (...). During earthquakes, cracks occur in the crust and in the mantle, and large masses get into motion (...) — an underground 'spring' allows the Earth to wobble", wrote the Russian geologist Abramowitsch Drujanow (1984, p. 32). He also points out that perhaps with the "underground rock capacitors, the rhythmic rock sequences" in the ocean floor can be explained, which then can be traced back to the interaction of currents from multi-layer electric fields. With mechanical deformation, also a different phenomenon (mostly) cannot be explained, which has already resulted in train accidents: During an earthquake, railway tracks are sometimes peculiar extremely twisted oddly enough.

Lightning's are observed in dust storms, hurricanes and in the course of volcanic eruptions. Underground thunderbolts are not only triggers and detonators of earthquakes, but *can* also be the actual cause: "If assumed that the contact area of radioactive grains is equal to 300 square kilometres, then the amounts of radioactive decomposition of the electric energy discharge in the result is up to 10^{16} erg (= 10^9 joules, HJZ). According to calculations by seismologists, this is the energy of a weak earthquake" (Drujanow, 1984, p. 34).

The entire fault zone at the Sumatra earthquake on 26 December 2004 consisted of 130,000 square kilometres and the released energy of 4.3 times 10^{18} joules was equivalent to a bomb of 100 billion tons of explosive power (Bilham, 2005).

Thus, this also explains that these electrical discharges lead to aftershocks in far afield areas. If the shells of the Earth's crust and mantle work like electrical capacitors, a static electric field develops in the dielectric that stores energy and produces a mechanical force between the conductors. In this way we find the cause of effects like the slowing of the Earth's rotation and especially the documented shift in the Earth's axis after the Sumatra

Fig. 49b: The Biefeld-Brown effect. The *schematic drawing illustrates the movement of a freely hung up condenser that is placed under tension. As soon as the condenser is under tension, a forward movement takes place*

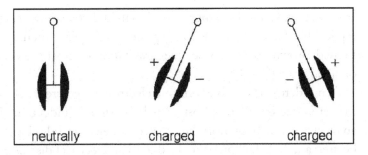

in *the direction of its positive pole. The author advances the following for discussion: If in its past the Earth operated as a sort of condenser, then it might be able to find a cause for the shift of the Earth's axis. While the dinosaurs were alive, the Earth's axis was still standing straight (ice-free Poles). Every type of known mechanical force, except for a planetary collision, is too weak to have put the large mass of the globe into a skewed motion.*

earthquake – without to take in account purely mechanical causes. In my book *Mistake Earth Science* is described the "Biefeld Brown effect", which was proven by Townsend Brown but not conventionally acclaimed, taking into account electric forces effects in the solar system like the tilting of planets axis without any mechanical forces, free of causative stress. This Biefeld Brown principle has been put up for discussion in case of the Earth as the "geo-capacitor theory" (Zillmer, 2007, p. 234 ff.). Therefore, to shift the geographic poles of our Earth there is required *no* mechanical but rather an electric force and thus tilting can occur quickly as we have seen during the Sumatra earthquake. The basis of this principle is a free-mounted condenser (Fig. 49b). One sets it under pressure, a forward movement in the direction of its negative pole is caused. A reversal of polarity also causes reversing the direction of motion and thus a swinging of the axis of the Earth

Fig. 49b: Electro gravity. The *Biefeld-Brown effect also causes a gravitation effect if the condenser is installed vertically on a bar scale. According to the situation of the positive pole, an anti-gravitational (unloading) or gravitational (loading) effect occurs. This effect represents a scientific anomaly according to conventional wisdom, since it also operates in a high vacuum. In contrast to the known effects of electromagnetism, by virtue of experiments by Townsend Brown a connection between electric charge and gravity was discovered.*

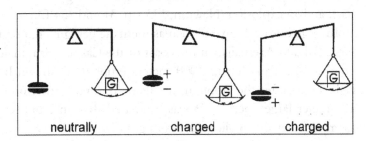

from one side to the other. The "Biefeld Brown effect" can be demonstrate and is replicable. There is a relationship between solar intensity and seismic activity: The Biefeld-Brown effect also causes a gravitation effect if the condenser is installed vertically on a bar scale (Fig. 49b).

"Thunderclouds with enormous electric charge create electric charges in the Earth of the same size but with opposite algebraic sign by induction. Their presence is a signal to an electrical Earth interior; the cause of an electrical discharge of our Earth (...). Through heating and liquefaction of rocks due to an electrical discharge can also explain the origin of volcanoes. Can the electric channel reaching deep into the interior be the beginning of an underground-lightning? The molten material expands and rises up under the influence of the lateral pressure. The lightning, however, can convert the stone material to ashes, crush them uniformly, heated and hurl over the Earth's surface in huge amounts" (Drujanow, 1984, p. 54 f.).

Such flashes in the lithosphere can leave diamond vents to be known as *explosion tubes*. These channels are testimonials of powerful discharges between the mantle and the Earth's surface which remain after the disruption of an underground capacitor.

The failure of the electrical equipment on the fishing boat *Bintang Purnama* during the Sumatra earthquake on the morning of 26 December 2004 could also be explained in this manner. The boat was leaving the road of Malacca, when it was hit by three violent waves and was raised up to 35 meters above the normal water level. Interesting is what was reported by the electronics: the instruments went to act up. The GPS navigator showed confused positions; the plotter, the sonar and the radio became inoperative.

Discharge processes also occur in the Venus atmosphere according to the latest research findings to the surprise of the experts. The *Venus Express* mission finally confirmed that there are lightning's. Bursts of electromagnetic waves were detected which had occurred by lightning's in the overlying ionosphere (Russell et al., 2008).

The electrical effects and interactions of the Earth – and also the other conductive celestial bodies – are an important, previously neglected factor. Another factor is the gas that emerges from the Earth. "The gas eruptions can either be triggered by earthquakes or can be the cause of this" (Gold, 1999, p. 143). Already Isaac Newton was of the same view as ancient writers that sulphurous gases are abundant in the Earth's interior and combine with minerals. Sometimes sulphurous gas catches fire, triggered by a sudden flash and an explosion (Newton, 1730, p. 31 and 354 f.).

In large outbreaks, gases such as methane get ignited, possibly by electrical discharges. Near Baku in Azerbaijan at the coast of the Caspian Sea, in an area with many large mud volcanoes, a flame up to 2000 meters erupted, as shown by an old photograph. Even eight hours after the eruption, the flame burned furthermore with lesser height out of a 120-meter large opening. It was estimated that an eruption of this magnitude requires approximately one million tonnes of gas (Photo 10).

Figure 50: Volcanic lightning. In the vicinity of volcanic eruptions, there are frequently discharges of lightning. See Photo 28 and for example:

1 Lightning streaks across the sky as lava flows from a volcano in Eyjafjallajokul, Iceland, April 17, 2010.
(REUTERS/Lucas Jackson)
2 Lightning bolts appear above and around the Chaiten volcano in southern Chile May 2, 2008. Cases of electrical storms breaking out directly above erupting volcanoes are well documented.
3 This photo is from the Cerro Negro in Nicaragua in 1971 (Decker / Decker, 1997, p. 56, Photo: Franco Penalba).

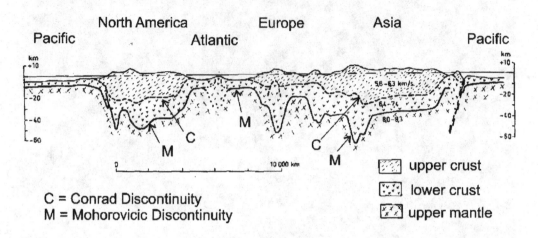

Pacific   North America   Europe   Asia   Pacific

Atlantic

C = Conrad Discontinuity
M = Mohorovicic Discontinuity

upper crust
lower crust
upper mantle

Fig. 51a: The Earth's crust. From the results of seismic measurements, the cutaway through the Earth's crust at 45 degrees of latitude north shows that the oceanic crust is distinguished by the absence of a granitic (upper) crust. Therefore, under the oceans there is no Conrad Discontinuity as a border zone between an acidic upper crust (granite) and a basic sub crust (basalt). The Moho Discontinuity probably represents a material border to the ultra-basic mantle lying under it. There could be a "Drainage Basin" (water basin) between both discontinuity zones (= Moho and Conrad discontinuity). The distinctive borders are additionally marked by a distinct increase in P-wave speed of 5.6–6.3 km per hour in the upper crust to 6.4–7.4 km per hour in the sub crust and finally to 8.0–8.3 in the mantle. Cutaway after Berckhemer (1968/1997).

From antiquity to the beginning of modern times, in the depths of Earth, it was thought that there was a water basin or a system of huge cavities filled with water (Tollmann, 1993, p. 148). Beneath the granite basement rock of the continents is to find the so-called *lower crust*, which consists of basalt and is situated above the upper mantle with mineral- and salt-bearing water – explained in my book *Mistake Earth Science* (Zillmer, 2001, p. 205 ff.).

This basaltic layer with salt water is a good electrical conductor and acts as capacitor, which is ideal to store electric charges respectively energy. When the capacitor is discharged, earthquakes occur in the crust and mantle pieces. Suddenly, a current flow is set in motion in the direction of the atmosphere and ionosphere by a discharge. The Earth's crust swings and huge cracks start to develop. Atmospheric lighting before most of the earthquakes is evidence of these electrical phenomena.

This scenario occurred mainly at weak points in the Earth's crust like the crack zones along the tectonic plate boundaries which are going to be open like a zipper as a result of explosive gas degassing processes and close thereafter. The electric power causes

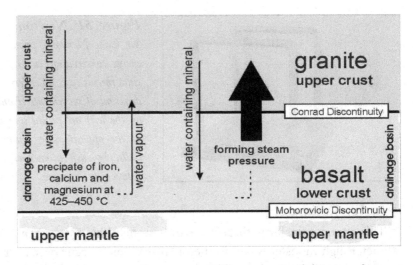

Fig. 51b: Drainage-Basin Theory. Water seeping through the lower crust, existing between the Conrad and the Mohorovicic Disconti-nuity (= lower crust layer), is heated. Due to the ruling temperatures water vapour is originated which establish *a steam pressure that puts pressure onto the upper crust. The mechanics of these procedures are displayed from left to right. If the pressure gets to high there will happen volcano eruptions like a lightning and also a slow expansion of the Earth crust. This is slowly but continuously going forward to slow down the pressure in the Drainage Basin.*

fluctuations in the Earth and even the Earth's axis can shift suddenly. At Christmas tsunami 2004 in the Indian Ocean, the shift was about eight centimetres. But the Earth's axis is anyway not fixed to the Earth en bloc. The North and south Poles move to about four inches per day. With the help of the electric model, this can be explained easily. In contrast to this, the conventional geophysical model requires rearrangements in the mass of the Earth's interior. But this is hardly to explain taking into account physical laws.

An explosive gas emission process accompanied by electrical discharges and earthquakes in the oceans, generally regarded as undersea earthquake, may be responsible for the occurrence of tsunamis. On the one hand, the ocean floor opens up, which could lead to large changes in volume and vertical displacements. On the other hand, the entire water column vibrates up to a height of several kilometres through the gas emission. This energy is transferred to the water molecules which are moved into vibration. The adjacent water molecules are pushed but without that these move away with the wave. This knock-on effect of energy transfers happens vertically over the entire height of the water column and is thus also transferred horizontally with high velocity. A wave of only low altitude is created lateral to the gas eruption because the water molecules are only stationary-rotating a little bit. In other words, no water is flowing, but only the kinetic energy is transferred from one water molecule to the neighbouring water molecule, which are themselves quasi-stationary. This is underlain the the well-known *elastic collision* principle (Fig. 50).

Figure 51: Newton's cradle. According to Sir Isaac Newton's principle, the momentum and energy conservation are related to elastic collisions, since in this case, no external forces contribute to the direction of movement. If one or more balls (shown on the left) are pulled and taken away, it vibrates at the opposite side against the same number of balls, in such a manner that the balls lying in between do not move.

This energy transfer is therefore hardly noticeable in the open sea, in contrast to the 30 meter high monster waves in the open sea which occur only near the water surface if storms and ocean currents arise. But tsunamis become giant waves only if they are hitting the continental shelf or impinge shallower coastal areas, so the kinetic energy collides with a barrier and is converted to potential energy, upheaving water molecules (Fig. 52). The alleged jumpy movements of tectonic plates centimeter by centimeter is not sufficient for the momentum, however, to generate a tsunami or to bring *all* molecules of a several kilometres high water column in the ocean to vibration and this can be experimentally shown on a smaller scale in a garden pond.

Figure 52: Tsunami. The energy from the momentum in case of shifting of two continental plates slightly against each other centimeter by centimeter is not at all sufficient to vibrate all the water molecules of a several kilometers high water column. With a gas degassing, however, energy is transferred over the whole water column so that the water molecules transfer the momentum in the form of an elastic collision (Fig. 51) to the next molecule and so on, up to the continental slope where a huge wave is formed for the first time.

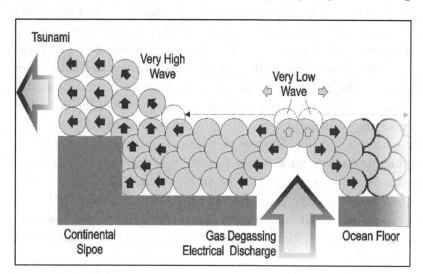

A prerequisite for tremendous tsunamis should be electrically conductive materials in the subsurface, which can be discharged and magnetic fields can be generated. Something of this occurrence was observed in our solar system. Unexpected changes in direction of the magnetic field were determined by the NASA's space probe Galileo at Jupiter's moon *Europa* in January 2000. At a meeting of the *American Geophysical Society* on December 16, 2000 in San Francisco, it was explained that the registered information exact corresponds to the data that would be delivered by celestial bodies with a shell of electrically conductive material; a circumstance for example which deliver a salt-water ocean inside of this moon under the crusty surface. Also inside Callisto, another moon of Jupiter, there is an electrically conductive shell presumed to generate a magnetic field, which can change its polarity. In the Earth, such an electrically conductive layer exists in the depths. It is the basaltic lower crust, in which is flowing salt-water and has already been discussed before as *Drainage Basin* (Fig. 51a and 51b). This layer in the depths of Earth should be to have the same function than those in Jupiter moons. The electrical currents, produced in the underground salt-water layer, are discharged from time to time and create magnetic fields accompanied by thunderbolts in the depths.

Structure Formation of the Earth's Crust

The sunspot activity reaches a maximum at an average of every eleven years, and during this time period also happens more earthquakes. Actually a connection between solar intensity and later on following seismic activity is detected (Drujanow, 1984, p. 34).

Minerals and rocks are often rhythmically banded, which was previously explained by gravity forces, sequential supply of materials, rhythmic precipitates and similar chemical-physical phenomena under pressure and temperature influence. Mechanical explanations alone are not able to interpret the observable diversity banded structures (see Fig. 53 and 54). But finely layered sediments (so-called *rhythmites*) can occur due to internal phenomena of self-organization processes (such as *Liesegang rings*) which can reinforce by external energy potentials. In addition to the gravitational field, the electric field of the rock shell of our Earth (lithosphere) is to be equipped with major structure-building forces, including the formation of deposits (Jacob / Krug, 1992). Through nature related electrolysis-experiments, Professor Karl-Heinz Jacob (Technical University of Berlin, Germany) could produce rhythmic mineral structures in the laboratory without any external forceful impact, only because he let flew a constant low-voltage current through an initially chaotic mixed "deposit". Therefore, wildly banded formations are able to build without any mechanical forces, only while an electrical current is flowing through.

For the formation of geological layers the flow of electricity or the operation of electric fields was not considered yet, although with geo-electrical research have been

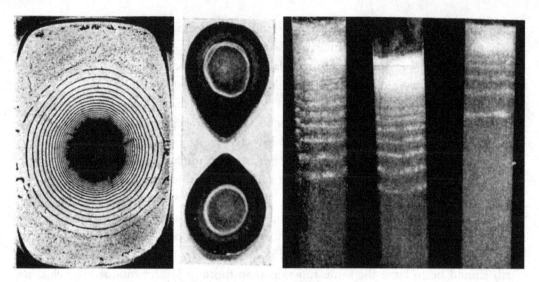

Figure 53: Self-organization. Left picture: The diffusion processes and rhythmic precipitates have a very special meaning in the geology, since such complex structures cannot be explained by mechanical modes of action. Certain chemicals are concentrated in the rings, while the spaces between are virtually free of such chemicals (Liesegang, 1913, p. 81). Middle picture: The apparent chemical remote action of two diffusion circles can be expressed through distances of several centimetres. But these have nothing to do with direct molecular forces, because such a range is calculated by physicists to a maximum of 0.0001 mm (ibid., p. 156 f.). Right picture: "Just as new parallel structure can subsequently occur in case of pressuring an already existing slaty medium, such stratification takes place here as a result of chemical processes also ... The electricity science term a related phenomenon, which sometimes occurs in a Geissler tube, as stratifying of a light pillar. And the same word is used by the histologist, when he speaks of structures in the bones, the naturalist, when he speaks of the mollusc shells. The geologist knows of the stratification process without mechanical forces but wants to explain such appearances in a geological medium as some type of sedimentation " (ibid., p. 85). The flow of an electric current enhances the process of the formation of rhythmic mineral structure through self-organization.

searched some mineral resources in the underground. Geologists typically not deny the underground electricity, however, but concede this principle an insignificant role.

Engineers know the electrical charging of cement, flour or coal dust during transport in pipes with compressed air. It can even build large electrical charges so that explode coal dust in pipelines or bunkers. Explosions and fires of coal dust and mine gas (methane) are feared. Traces of destruction were observed on boulders which look alike to the electric charges in non-conductive materials (dielectrics). With the destruction of crystals, loads occur at the fractures, even in ordinary table salt. Tiny flashes spring between the breakages and radio waves are broadcasted.

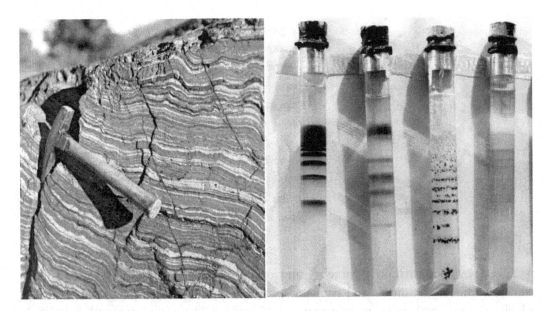

Figure 54: layering. **Left picture:** *According to geological interpretation, this layering has occurred from an ore deposit on Lake Superior (USA) in which silicic acid and ferric oxide rich sediments were deposited alternately constantly over long time periods almost imperceptibly slow layer by layer (Schidlowski, 1988, p. 187). Self-organization provides an alternative explanation model without mechanical sedimentation processes).* **Right picture:** *This is a picture which took Donald B. Siano (2006) 20 years ago of a series of experiments on Liesegang rings – with the eyes of geology apparently processes of sedimentation showing.*

Through electrical discharges, bullet-type traces are left behind. Thus, perhaps the observations of geologists I. W. Muschketow (Drujanow, 1984, p. 29) can be explained: "The cracking of the cable Kerkyra on 7 December 1883 took place in the sea depth of 100 meters. The marine telescope clearly detected a fault line on the soft limestone soil ... and round star–like shaped openings in the limestone, similar to the shatter of glass by a bullet ..." Such bullet–like traces can leave also electrical discharges, which melted and marked basaltic bodies, also tuffs consisting of volcanic ash, and hit quartz grain.

Quartz–like minerals are often found in the Earth's interior. The granites of the continental shelf consist largely of quartz (SiO_2), a salt of silicic acid, which is chemically precipitated out in the lower crust respectively *drainage basin* and then will be transported with the water vapour upside through the seismic zone border, the so-called Conrad discontinuity, and it is enriched in the upper crust (see Fig. 51b, p. 125 and in detail in: *Mistake Earth Science*, 2007, p. 205 ff.). If one exerts pressure on those crystals, mechanical energy changes directly into electrical energy. This piezo–electric effect can also be observed on crystals of our daily use, like in cane sugar. In the Earth's crust, crystals are

Figure 54a: Geodes. They are geological rock formations, essentially rock cavities or geode with internal crystal formations or concentric banding. Geodes occur in sedimentary and certain volcanic rocks. The exterior of the most common geodes is generally limestone or a related rock, while the interior contains quartz crystals and/or chalcedony deposits. Other geodes are completely filled with crystal, being solid all the way through. These types of geodes are called nodules. This geode have been halved and polished. Is to see here a melt nest? Is the cause a piezo-electrical discharge?

polarized by deformation, whereby previously electrically neutral structural cells constitute an electric dipole and with this procedure, an electrical voltage is occurring, which can be discharged as underground lightning. These underground lightning's steered the geological evolution of the Earth.

"With the gradual accumulation of heat and substance, the expansion stress in the crust is increased and discharged periodically by the active phase of earthquakes and volcanic eruptions. Therefore the Earth does not grow evenly, but it grows in a pulsating manner, which also reflects the decrease in its rotational velocity "(Oesterle, 1997, p. 86).

The geologists believe that the molten and hardened edges of the basalt formations can be explained as traces of high pressures and temperatures. But, the same traces

Figure 54b: Fulgurites. Left picture: A lightning tube respectively fulgurite (topview and front view) from Okechoobee in Florida (Mario Hendriks, 2006). Middle picture: As a fulgurite tube is amorphous it is classified as a mineraloid and remembers a geode with a melt nest looking interior (Fig.

54a) Right picture: Fulgurites in a sidewalk of Minneapolis at Colfax Avenue and W. 24th Street created by lightning. The scar is immediately below an ordinary city power line pole, and may be it is possibility that the scar was created by a downed power line.

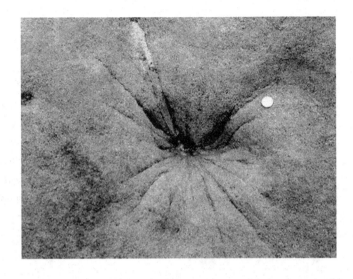

Figure 54c: lightning struck
The somewhat sphincter like an impact spot shows where lightning struck the ground – beneath the center there's likely a fulgurite.

can be discovered *in the interior* of basaltic masses, even in the form of thin veins, as if a burst of fire would have created a chaotic pattern in the dark rocks. Melt nests are also observed regularly. How does energy go into such a melt nest, if even the smallest sign of energy is not to be seen around? Does the ignition spark come from inside the Earth? On rock samples are to be found traces of destruction, which are similar to those caused by electric charges in non–conducting materials (dielectrics), writes Drujanow (1984, p. 36).

Not only melt nests occur in rocks through electrical discharges, but also hollow flash tubes, so–called fulgurite (from *Latin*: fulgur = thunderbolt). Fulgurites are formed in quartzous sand, or silica, or soil by lightning strikes over a period of around one second, and leaves evidence of the lightning path and its dispersion over the surface. Fulgurites can also be produced when a high voltage electrical distribution network breaks and the lines fall onto a conductive surface with sand beneath. They are sometimes referred to as *petrified lightning*. The glass formed is called lechatelierite which may also be formed by volcanic explosions and meteorite impacts. Fulgurites occasionally form as glazing on solid rocks and they can occur as far as 15 meters below the surface that was struck and the tubes can be up to several centimeters in diameter and five meters in length. Their colour varies depending on the composition of the sand they formed in, ranging from black or tan to green or a translucent white.

4 The Earth is Loaded

In recent times, in their observations astronomers reveal more and more contradictions to the scientific consensus, which is presented over several decades, under which the universe might have been come into being with a huge explosion 15 to 20 billion years ago today, and since then, it has been continuously expanding. Cosmologists, however, have trouble dealing with these contradictions and to turn inside out their model of the world in which we live, wrote Hans-Joerg Fahr (1992), professor of astrophysics at the University of Bonn.

Neutral spherical shell of Earth

In order to form a stable solar system, the Sun must be in static equilibrium (Fig. 43, p. 104). Therefore, positive and negative forces (electricity) must necessarily are in balance in the *neutral spherical shell*. In case of the Earth this sphere is not visible in contrast to the Sun; because this statically neutral zone is located *below* the planet's surface. This also explains that the electrical conductivity accelerates about 100 kilometres deep into the Earth and then decreases again.

Therefore, the Earth was never completely melted, but magma is mainly formed in the area of the *neutral spherical shell* and some surface near layers. A three-dimensional model developed at the Harvard University suggests a mixing of the upper mantle without fixed or systematic up and down streaming zones (so-called *convection cells*). These studies made by Don Anderson and Adam Dziewonski (1988) with the help of seismic tomography showed that the hot layers decrease *with increasing depth*. Already at a depth of just 550 kilometers, there are only very few hot areas and no *very* hot areas exist anymore — contrary to the geophysical theory of rotating convection cells in the mantle with an upward–streaming hot and downwards streaming cold material. The model of plate tectonics and the therewith unalterable firmly connected *convection cells* are definitely more than called into question, especially since Anderson and Dziewonski found that hot material is directed to the spreading zones more horizontally from the side, but not from below, and only "by a few, widespread thermal anomalies is spring-fed" (Anderson / Dziewonski, 1988, p. 77). Therefore, magma *cannot* result from a melt-liquid state of the Earth's deep interior, and it can also not be rising from a great depth as a convectional gush below the mid-ocean ridge respectively spreading zones and then to flow off in both directions horizontally, thus shifting the continents (Fig. 6, p. 20).

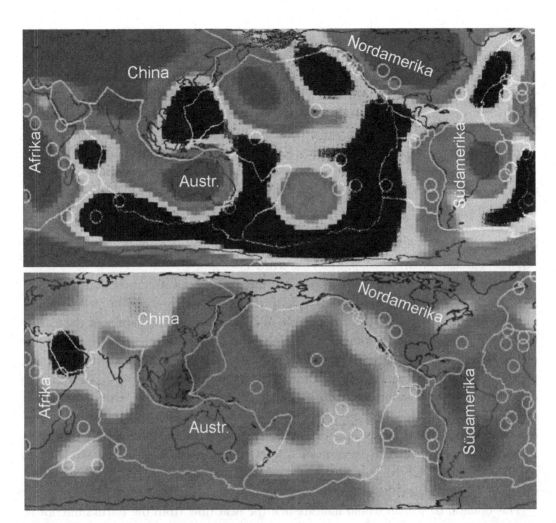

Figure 55: Temperature distribution. *The tomographic maps by John H. Woodhouse (Anderson, Dziewonski, 1988, p. 72 f.) show black-coloured areas with hot zones that cross over to light grey areas in cold regions (dark gray). The picture above shows the conditions at 150 km depth with several hot zones and the lower tomographic map of 550 km depth shows only one single hot zone which is located in the Red Sea. The bottom zone of the upper mantle consists almost entirely of cooler material at least in the spreading zones, where hotter material should exist according to the plate tonic theory.*

New investigations of the super-eruption in the Yellowstone area 640 000 years before today (on the geological time scale) ensured a solid surprise, because the magma, contrary to expert opinion, *did not* come from an emptied magma chamber which gets filled up again and again with hot material. Yellowstone volcanism became an independent "source", which was located *near* the surface of the Earth (Bindemann, 2006, p. 43–44).

The model presented in this book of the *neutral spherical shell* of the Earth's crust can explain why in this sphere exists hot material and *not in deeper regions*, and certainly not much deeper in the interior of the Earth. The *neutral spherical shell* is thus a heated zone: in the Sun as well as in the Earth. The temperature in the heated zone must decrease in both ways, outwards to the surface as also towards the core of conductive celestial bodies. The Earth should be as cold as the Sun in its core. In other words, in today's Earth's core, there is the static potential energy, and externally on the periphery join the *natural spherical shell* in the form of a heated zone with unevenly distributed hot-fluid matter. Overlying this, a solid state has evolved (crust is formed) over millions of years.

The tomographic maps (Anderson / Dziewonski, 1988) show (Fig. 55, p. 133): The deeper the material exists in the mantle below the zone with heat anomalies (= natural spherical shell!) the colder it is going to be. For this situation the well-known German physicist Pascual Jordan (1966, p. 87 f.) has written:

Already "MacDonald has pointed out that the sonic speed inside the Earth show a fact which looks physically disconcerting from the viewpoint of the conventional interpretation in case of monotonic increase of temperature in the depths of the Earth. Only at a depth of nearly 200 kilometres, the sonic speeds decrease as it would be expected, but only if they were not to be influenced by the increase of pressure and density. Thereafter, however, they increase a lot – to an extent that is physical surprising; even though it is to admit that the density increase result in an increase of the sonic speed, especially of the longitude waves. But it seems strange that there is such a strong overcompensation of sonic speed decrease caused by temperature increase resulting from a pressure increase which is accompanied by a sound speed increase, as MacDonald rightly points out. This paradox disappears if we consider it possible that temperatures of the Earth's interior are rather low than high."

This opinion by Jordan would indicate that the heat flow from the Earth's interior is generated not by radioactive decomposition, but by chemical energy. This is most likely because much lower activation energy is required for the formation of a molecule and the separated amounts of energy are lower. "The chemical energy as an important source of heat in the Earth has been largely ignored, but it could well be true, especially in terms of production of movement" (Gold, 1987, p. 186).

If rules a balance between the potential energy in the Earth's interior and the external kinetic energy in the dynamic *neutral spherical shell*, then electrical discharges will result inevitably and earthquakes, volcanic activities and lightning's will occur necessarily. Thunderstorms in the atmosphere are therefore a visible manifestation of a discharge, which shoots out from the Earth's crust into the atmosphere. Here, the ionization of the air is regulated primarily by two factors: On the one hand the galactic cosmic radiation (Volland, 1991, p. 286) and on the other hand the radioactive materials to be contained in the crustal rocks (for example uranium and thorium).

The noble gas radon, one of the decay products of uranium, diffuses into the atmosphere, where it is rapidly converted in polonium with the release of alpha radiation. A portion of the air molecules is ionized by the high-energy alpha radiation. "The so–formed ions and electrons combine nevertheless very quickly over a hydration process with water molecules and in this way occur positively- and negatively charged so-called *small-ions*. These *small-ions* can be further deposited on aerosol particles (small airborne particles, such as tiny water droplets, dust, smoke, etc.), whereby long-lived *big-ions* are formed, or they can be recombined" (Volland, 1991, p. 285 f.).

Suchlike aerosol particles affect the climate significantly because the air temperature decreases with the increase of aerosol particles, such it has shown after volcanic eruptions: The climate becomes colder. Humans also produce masses of aerosol particles, but this is going to be neglected in the calculations of climatologists, and recorded at an insignificant value of *only* one watt per square meter (Rahmstorf / Schnellnhuber, 2006, p. 45). The mankind air contamination leads not to a heating, but actually to a cooling. Temperature decrease factors are not encouraged, however, in today's climate debates, and there is a strong need to clarify why since 2000, it has not become warmer of the mean annual temperature, despite increasing carbon dioxide concentration.

Climatologists and politicians like to keep secret the fact, for example, that the temperature series of measurements in Karlsruhe, Germany, show that the 1820s were similar as warm as the 1990s and 1822 was the warmest year in the past 200 years! There are very few reliable temperature records that extend so far back in time. Modern temperature series of measurements begin just about 100 years ago in the midst of a cold phase, the final phase of the Little Ice Age. With January, March, May, October, November and December in Karlsruhe were measured six of the warmest months in the past 200 years in the first half of the 19th century until the year 1852, before a cold period began in which the modern temperature measurements principal started.

As the Sun at the beginning of the 21st Century is quite unusually inactive for a long period, respectively is exposing a minimum of sunspots, it should going to be colder rather than warmer for longer periods of time. In such a manner, the climate continued likewise during the Maunder Minimum, the coldest period of the Little Ice Age, with greatly reduced sunspot activity in the second half of the 17th Century. According to the temperature series from Central England, temperatures then rose at the beginning of the 18th Century up to nearly two degrees Celsius in just 40 years starting 1700 and ending about 1740.

Even earlier, before the Little Ice Age starts, it was in such a manner dry in 1130, that one (once again) could wade through the Rhine near Cologne in Germany. From December 1186 to January 1187, the trees were in bloom in the winter time and there was aptly talked about that the summer has replaced the winter in Cologne (Glaser,

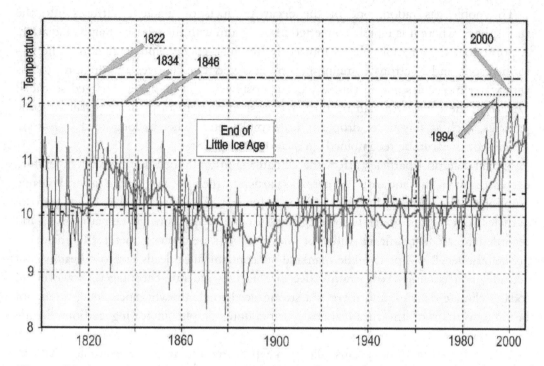

Figure 56: Average yearly temperatures in Karlsruhe from 1800 to 2008. 1822 was the warmest year of the last 200 years. Since 2000 it has not become warmer until 20012 (up to the publication of this book).

2001, p. 61). Then, in the year 1342 occurred the worst natural catastrophe of the last 1000 years in Central Europe because a millennium flood destroyed many large bridges and gorges were torn up to 14 feet deep (ibid., p. 66).

Always earlier, in the 11th Century, vineyards were cultivated in England, and Vikings grow wheat and grain and carried on livestock production on the lush island of Greenland – described in detail in my German book "Columbus came too late". Not humans, but the Sun controls the Earth's climate significantly, because in consideration of five months of darkness it would not be able to grow enough food for many cattle's.

Solar Energy Supply

The celestial bodies in our solar system are connected over the interplanetary magnetic field with each other and with the Sun. When this magnetic field is directed away from the Sun, then in the case of reconnection, thus "the new connection of the field lines

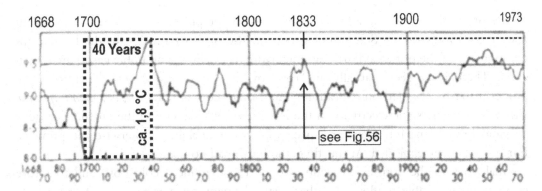

Figure 57: Temperature series from Central England. *This ten-year-average temperature profile since 1659 shows a steep rise in temperatures from the ice-cold "Maunder Minimum" onward to 1740 of nearly two degrees Celsius in 40 years. The temperature high range was around 1830 until 1840 as already noticed for Karlsruhe (Fig. 56) and the minimum temperature was around 1890, the starting time of modern temperature measurements. Source: According to Manley, 1974.*

coming from the Sun should be connected to the Earth's field of the northern polar cap, while the south polar cap is connected with magnetic field lines going out into interplanetary space. In this case, only in the northern polar cap cosmic rays coming from the Sun should penetrate", notes Dr. Manfred Scholer (1991, p. 128), colleague at the *Max-Planck-Institute*, Germany.

The planets and other electrically conductive celestial bodies like comets are powered by the Sun, and the cosmic energy (Sun electricity) can penetrate through a pole and is stored in the interior of planets – like in our Earth. Excess electricity can pour out, because this is the reason why gas planets emit more energy than they receive on their *surface* in the form of sunlight. "The largest source of particle radiation emitted by planets builds *Jupiter's energetic electrons* which can be verified over large areas of the inner solar system (Wibberenz, 1991, p. 48). In case of inner planets with solid crust, the situation is somewhat otherwise, since they have a relatively electrically neutral planet surface.

The cosmic energy flowing through a pole is swirled in the outer core, and with the formation of a point-like eddy (*irrotational vortexes*) with *closed* field lines there are generated elementary particles of matter *and* also heat is released. These processes proceed at relatively "cold" and not at hot temperature. The *irrotational vortexes* combine to form nuclides (atomic nuclei), atoms and molecules, which in turn also represent *irrotational vortexes* showing the same characteristics. The more complicated a nuclide is the lower the probability of its formation. Therefore, accumulation of simple and stable nuclides takes place mainly from hydrogen and helium.

A significant influence on the processes of planets emergence carries out the distance of the proto-planets to the Sun. In case of the inner planets, as a result of turbulent energy discharges of the Sun reaching beyond the orbit of the Earth, the lightweight components of the primeval Earth (atomic hydrogen, helium, and noble gases) are blown away. These have been partially absorbed by the outer gas planets. Therefore, the primeval Earth consisted only of an inner and outer core and a shell, the *neutral spherical shell*, in which the static electricity in the interior of the planet and the kinetic electricity flowing through the interplanetary space was in equilibrium – otherwise the primeval Earth would have exploded. At this time, the Earth had no mantle.

Analogous to the Sun, in the centre of the Earth reside a cold pole, and therefore cosmic energy is able to flow into the Earth. At very low temperatures, however, no chemical processes take place, whereas in (normal) low temperatures, the formation of atoms is more common. At high temperatures, however, outweigh the destruction of atomic bonds, with energy consumption and decrease in temperature. At very high temperatures are acting no chemical processes, because all the atomic bonds are broken. Therefore, chemical processes take place in a cold outer core, at low temperatures.

"Studies show that most chemical transformations happen at moderate temperatures, approximately in the range between 0 and 100 degrees Celsius" (Oesterle, 1997, p. 69). Using thermodynamics (Oesterle, 1990) or chemical kinetics, it can be showed that the most probable temperature of chemical change is exactly equivalent to that of the human body temperature, thus nearly 37 degrees Celsius. This most probable temperature of chemical changes has been calculated *without* taking into account the effect of pressure on the chemical change. This dependence is not linear, and there is a borderline temperature, which is 1600 degrees Celsius for the formation of minerals (Oesterle, 1997, p. 71 ff..).

At least now, one can point out that the outer core is proven in a molten state and therefore could hardly to have formed there matter from energy. As evidence should be assumed that the outer core passes on only

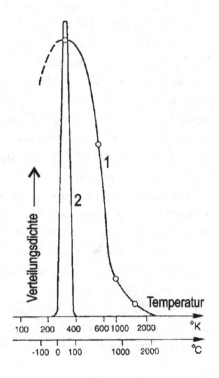

Figure 58: Development. The statistical distribution (frequency) of chemical compounds according to their formation-temperatures for minerals (1) and organic compounds (2). According to Oesterle, 1997.

138

longitudinal waves, but no transverse waves (secondary waves). Pascual Jordan (1966, p. 87) replied:

"But in reality we don't know whether longitudinal waves are able to travel into the outer core; what we know is simply that only one of the two types of waves – longitudinal or transverse penetrate the outer core. Now, in the exchange with Joel E. Fisher, the physician Frankenberger has examined with the help of detailed calculations, how seismic waves would travel from the mantle into the outer core, if we make the assumption (which seems to make sense physically) that the outer core has a much lower compressibility and thus feature a much larger longitudinal sound speed than the mantle rock: This shows that in the core – now also assumed to be in solid state – practically only transverse waves could enter."

We know even less about the inner core of the Earth. It is believed that the inner core is in the solid state, but the shear-wave speeds are derived only from indirect estimations (Berckhemer, 1997, p. 110).

As discussed earlier, Jordan (1966, p. 88) had determined that inside the Earth, the depth distribution of the sonic speed rather vouch for a statement of low and not of high temperatures, even though usually exactly the opposite is claimed. These low temperatures are possible within the Earth if the core of the Earth consists of metallic hydrogen. The inner core could, therefore, consist of frozen hydrogen, at a temperature of maximum 14.02 Kelvin (= –259.13 Degrees Celsius), which is close to the absolute zero temperature. In this form, the hydrogen forms a crystalline solid body. In which phase the solid hydrogen in the inner core is actually present, is a speculation, because experimental tests are technically difficult.

More important for our consideration with respect to its chemical properties is the outer core that lies over it, which should be in glow-molten state. At temperatures from 14.02 to 20.27 Kelvin, hydrogen becomes liquid and under very high pressure, it is metal. Also, the inside of the gas planets of our solar system and even some exoplanets consist of metallic hydrogen according to the latest research results. Under extreme pressure, metallic hydrogen is formed from atomic hydrogen and it gets an electrically conductive property. Only a few experimental data are available about this aggregate phase, because the production in a laboratory is extremely difficult and this condition is very short-living. But it is possible to produce metallic hydrogen. In fact, liquid metallic hydrogen is no fluid, but it is plasma containing only independent charge carriers. It is suspected for a long time that metallic hydrogen at high temperatures is superconductive. Therefore, it is possible that in the outer core is flowing electricity almost with no resistance, whereby the energy is coming from the Sun with.

Hydrogen is the simplest of all atoms, but at the same time, it *does not* represent the simplest form of solids or liquids. In 2004, a topological analysis of a projected state of liquid-metallic hydrogen was published in the journal "Nature" (vol. 431, 7 Oct 2004,

p. 666 ff.). Specifically, it is to show that liquid-metallic hydrogen may not only be exclusively either superconducting or super fluid, but it shows a new type of quantum fluid. The researchers assume that in the presence of a magnetic field, liquid metallic hydrogen can take several transitional phases from a superconductor to a superfluid (Babaev, 2004). It is also assumed that liquid-metallic hydrogen displays previously unknown electromagnetically properties.

In addition to metallic hydrogen also helium and traces of other elements are present in the outer core. Low temperatures, as is known from superconductivity, benefit eddy-currents and magnetic fields. The electrically conductive properties of the outer core are suitable to maintain cosmic energy in the form of electrical currents in the Earth's interior. Since each magnetic field always has an electric current as the cause, the magnetic field of the Earth can be explained as generated by electrical currents. Although physicists have debated for more than 100 years this simple method of generating the geomagnetic field is not accepted officially, because the Earth was and is viewed as an isolated celestial body, only influenced by gravitational force, and in this case no reason was found for maintaining the electric currents inside the Earth.

The electric and magnetic fields are perpendicular to each other, if a field is an open field line. In the case of the planets, it is the magnetic field that is developed perpendicular to the electric field lines. This dynamo is set in motion by the Sun and kept going on. On the other hand, according to the geophysical view the geo-dynamo should work in an independent manner, therefore only connected to the Earth. But it is not even explained the starting mechanism and how should the geo-dynamo remain in motion after the starting process? Supposedly three energy sources are available (Berckhemer, 1997, p. 131):

1) The heat storage of the core.
2) The latent crystallised warmth.
3) The gravitational energy during the freezing and sagging of the nickel-iron.

The first source is pure speculation, because it is possible that the core is alternatively cold and the actual heat storage is unknown. The second option is indeed qualitative feasible, but it is clearly too weak for the necessary dynamo drive. The third source is again speculation, because one does not know what material the core is made up of. Moreover, the Coriolis Force, by which the convection currents in the Earth's interior are diverted as a result of their own inertia and are compelled on a helical path, is too less: "The emergence of vortexes as a result of the Coriolis Force with due regard to the great viscosity of the molten mass in Earth core (huge pressure) and low temperature gradient (high conductivity of the substance) seems to be impossible" (Oesterle, 1997, p. 98).

Consequently, the origin of the magnetic field is questioned according to the conventional geophysical model, because the Coriolis Force should also swirl the field lines in addition to the convection currents, in order to increase the magnetic field strength thereby. Although when this effect occurs related to the field lines, it is too small to be responsible for a significant amount of the magnetic field strength. Other disadvantages of the dynamo model are discussed in the journal "Nature" (Backus, 1995).

If one replaces the word "iron melt" in the conventional model by "liquid-metallic hydrogen", then one has a cause for the launch mechanism and the maintenance of the magnetic field in this case of a *cold, electrical Earth with energy supply from the Sun.* The vortex effect of the electric flows in the outer core results in closing of the electric field lines, through which, as already described, atoms and molecules emerge. The thus materialized energy is enriched as matter by the rotation at the edge of the outer core and forms a shell, 100–500 kilometers, mostly 200–250 kilometers thick. This is the so-called but rarely mentioned *D″ layer*, lying at the *Gutenberg Discontinuity* or *core-mantle discontinuity* i.e. core-mantle boundary (Fig. 59, p. 142). Since matter is formed in the cold outer core, the *D″ layer* should consist of cooler, denser material. This was confirmed by seismic measurements and therefore leads to a paradox in the conventional geophysical model.

Since the cause is unknown for the appearance of the *D″ layer* (see Lay / Garnero, 2004), which is beside this also *much colder* than the surrounding mantle rock, it is presumed that we are talking here about the *D″ layer* that built irregularly and heterogeneously as deposits of subduction zones (Vogel, 1994). This idea has arisen from the hot-Earth theory, because "*cooler*" material cannot come from the outer core, if that should be hot 2900 degrees Celsius. Puzzles are not created by nature but by wrong conceptual models! How would the cooler, but particularly lighter ocean crust "drop from a plate like viscous honey" (Hutko, 2006) and reach through the denser material of the mantle till a depth of 2000 kilometers? But even if in the lower mantle, the oceanic crust is transformed in a high-pressure modification of quartz, then its density can transformed up only to $4.34 \, g/cm^3$, compared with $5.7 \, g/cm^3$ of the lower mantle material. Buoyancy and not immersion would be the appropriate model of thinking or in other words, the allegedly subducted ocean crust cannot drop from a plate like viscously honey and dive to the core boundary passing the complete mantle of the Earth (see Figure 7, p. 22).

Conclusion: This denser and cooler material of the *D″ layer* definitely does not consist of remains of subducted oceanic crust. It remains the only solution, that the *D″ layer* material is produced within the Earth, apparently in the relatively "cold" outer core. The cool-dens material of the mysterious *D″ layer* lying on the core-mantle boundary, should therefore *not* be originated from a downward movement of the Earth's crust; but instead of this, according to the cold-electric Earth model, this *D″ material is formed in the outer core and deposited unevenly at the edge of the core under the mantle.*

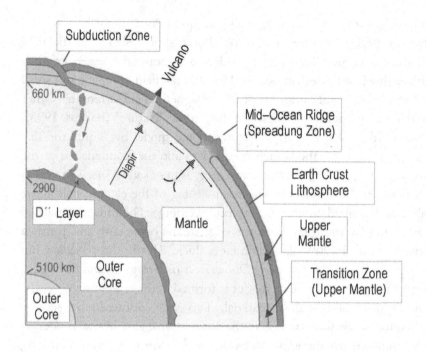

Figure 59: D" layer.

The material of the D" layer is cooler than the mantle material and separates the mantle from the outer core. One hypothesis is that the material of the D" layer is subducted ocean crust which dives to the core-mantle boundary through the whole mantle — although the ocean crust material is lighter than the material of the lower mantle (see also Fig. 7, p. 22). At the same time, as per conventional thinking, single upward-streaming diapirs in the mantle should come from the core-mantle boundary, and for example, provide volcanoes with hot material. But hot magma is found only up to depths of about 400 kilometers (Fig. 55, p. 133).

Therefore, the boundary between core and mantle shows no discontinuity of the chemical composition according to conventional model, but it is rather a mere phase boundary, because the pressure decreases with increasing of the distance from the centre of the Earth to the outside. The primeval mantle then grew steadily, without any input of matter: Thus 100 cubic centimeters outer core volume was transformed to 178 cubic centimeters of the (lower) mantle, with no increase in mass and in the weight. This assumption is based on the volume increase determined by density ratios in the mantle and in the outer core, proved by seismic measurements.

With the continuous conversion of energy into matter in the Earth's interior and the subsequent increase in volume growth, the mantle of the Earth grew in the radial direction steadily. At the same time, the average atomic weight of matter decreases with increasing distance from the centre of the Earth continuously, as a function of steadily outward decreasing pressure. If minerals go from the high-pressure phase to a low-pressure phase resulting from the state of decompression, then zones of phase transformation are occurring. These zones are misinterpreted as boundaries of different chemical composition in the depths of the Earth (zones of discontinuity). But actually, in each of these zones a volume increase takes place due to pressure-relief, by conversion of the same material only in an especially variant of lighter or less dense material modifications. The result is that the volume of the Earth increases in these boundary zones without increase in mass and thus weight.

Pascual Jordan (1966, p. 74) confirmed, as already impressively was shown in 1941, that the majority of boundary zones in the Earth's interior are phase boundaries rather than discontinuities in the chemical composition. There is no doubt about this, because the age of the Earth is not sufficient to produce a multistep chemical decomposition with separation of the material into zones of different composition, as stated in the German professional journal *Geologische Rundschau* (vol. 32, 1941, p. 215). Therefore, the modern interpretation of geophysics regarding the structure of the Earth is definitely wrong!

With the formation of substances in the outer core, the new atoms rise from inside to outside through the mantle towards the surface and conjoin this atoms to water, natural gas and oil on the upward way at particular depths in which certain pressure and heat conditions prevail; as will be discussed later on in detail.

But these processes are not limited to chemical processes under different pressure and temperature conditions, but – based on the observations of Pascual Jordan (1966) and Thomas Gold (i.e. 1992) – we need *additional* to take into account the electrical effects. Resulting from the diffusion of electrons outgoing from the interior of the Earth (Thomson effect), an electric field occur at the Earth's surface (Oesterle / Jacob, 1994).

"One can even say that the Earth's interior is a giant magneto-hydrodynamics generator which produces electricity. The same result is achieved with a temperature drop when a conductor is heated at one end and cooled at the other end. Each electrolyte is a conductor, and nothing is easier than to get into a hot-cold situation in the Earth's interior" (Drujanow, 1984, p. 52).

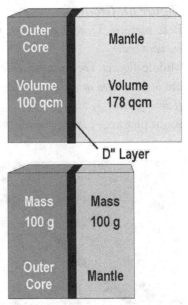

Figure 60: Volume increase. *At the core-mantle boundary (D" layer, Fig. 59) an increase in volume takes place as a phase transformation resulting from a pressure-release, while the mass remains constant. Even at further discontinuity zones towards the outside, more phase transformations happen with an increase of volume; notably in 410, 520 and specially 660 km depth at the boundary layer between the transition zone i.e. upper and the lower mantle.*

Depending upon pressure and depth for the optimum temperature, a process called self-organization of chemical elements takes place in the Earth. With the gradual accumulation of heat and substance, the extent of pressure in the crust increases, this is discharged periodically through the active phase of earthquakes and volcanic eruptions. That is why the Earth does not grow evenly, but it grows vibrantly which also reflects in the decrease in its rotational velocity (Oesterle, 1997, p. 86).

With the currently valid theory of the evolution of the Earth, it cannot be explained how, in the depths of the Earth, the explosion capacity of the rocks is still preserved in spite of activity over billions of years. "In reality it is not enough to change our concept of the Earth in a revolutionary manner. Instead our view of the cosmos must be revolutionary changed in the sense of Dirac's hypothesis" (Jordan, 1966, p. 107) – this is based on an ever-decreasing gravity. In the electrical model which is presented in this book, Dirac's hypothesis corresponds to a decrease in the cosmic energy and electricity. The inevitable consequence is that an expansion of the planets, but also of the Sun takes place.

If we consider the moon in this respect, it is to notice immediately that there are many grooves on the lunar surface. These fissures and cracks are an unexplained phenomenon. As on Mars, some of these grooves look like chains of small craters, others don't. But the numerous narrow fissures and cracks of the lunar surface are partly filled with craters and partly not. Because the grooves are running through the crater walls, the

Figure 61: Grooves.
Left picture: *Trenches on Mars, photographed by the probe Mars Odyssey. Enlargement A shows outward crater.*
Middle figure: *The moon has similar grooves. In these trenches as weakness zones in the crust, which can be interpreted as cracks resulting from an expansion of the moon, craters are formed in several areas (Lunar Orbiter 5, 1967).*
Right picture: *A lunar groove without Crater (NASA alleged Apollo 10, 1969).*

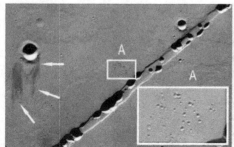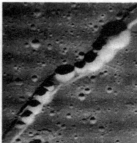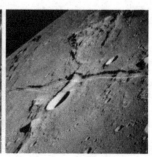

craters should be a secondary phenomenon on the one hand, because here in the area of the cracks, gas emission processes are physically able to occur more easily. On the other hand, a punctual crater formation can also be an impulse to start grooving. The view of all the grooves "as a single phenomenon of the lunar surface" suggests that they are a result of a (small) expansion, in the sense of Dirac's hypothesis, i.e. of a reduction of the gravitational force (Jordan, 1966, p. 38 f.).

Pace of Expansion

The growth of the Earth can be accelerated by an absorption process, in addition to those causes which are outlined here in this book. This process was introduced for a discussion by Professor Konstantin Meyl (University Furtwangen in Germany). Therefore should be absorbed solar neutrinos (*electrically neutral* charged elementary particles) *inside* the Earth, which has been demonstrated in principle as well. But even penetrate some 70 billion neutrinos per second through a square centimeter of the Earth's surface into the interior of the Earth there will be absorbed only a few neutrinos from the Earth – according to current knowledge. So there is definitely an increase in mass, but this is much too low to explain an expanding Earth. According to Professor Gerhard Bruhn (University of Darmstadt, Germany), it accounts for less than 8–10 mm per year – at least. It would be to examine whether solar neutrinos may be slowed down in the neutral spherical shell and / or in the cold electrical Earth's and will materialize in order to make way for a certain amount of *additional* mass increase of the Earth.

Regardless of this scenario, which boosts the "normal" expansion, the growth of the Earth took place in three phases. The first phase of an expanding Earth (without mantle) happened very quickly at cyber-speed because the energy– and pressure down grade in the interplanetary space was very huge. According to physical laws, energies and particles were flowing like water toward relative cold temperature points. One of these points was the future centre of the Earth. In the second phase of expansion the mantle started to form a crust until the Earth was enwrapped by a crust with a thickness which is equivalent to the present continental shelves. At this point, the Earth's surface represents all present continents including today's continental shelfs. These entire crust surrounds a smaller Earth completely (Pangaea-Earth). The diameter of this small Earth is calculated at somewhat more than 60%, but may be 62 or up to 65% of its current diameter, taking into accounts a tolerance at the continental plate's edges. In other words, we can place all present continents on this small globe as a completely enveloping crust.

The time period at which the crust of the Pangaea-Earth began to break apart as a result of the expansion due to high tensile pressures in the crust, has been fixed at an age of 200 million years – according to the plate tectonic timescale.

Figure 62: The primitive mantle.

The smaller primal Earth (left) had an undivided primitive mantle, built by phase transformation that grew more and more which resulted from pressure discharge, while the outer core became smaller. Therefore the Earth expanded more and more. As a result of pressure discharge, there was a further differentiation of the primitive mantle with a corresponding build-up

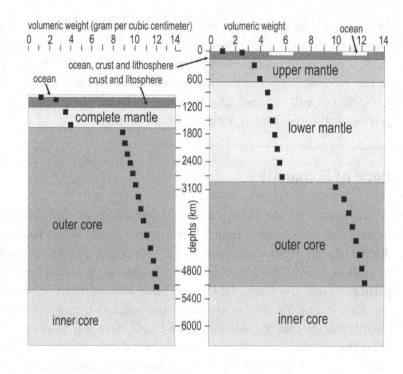

of several phase boundaries (right image). New after Pickford, 2003. Compare figure 59.

The data of the growth and the expansion rate of the Earth range from 0.5 millimeters (mm) per year after L. Egyed (*Geol. Rundschau*, 1957, vol. 46, p. 108 and 1960, vol. 50, p. 251) to about ten times of this, that is 5 millimetres per year (Jordan, 1966, p. 77) and up to 22 mm for the diameter, that is 11 mm of growth per year on average for various points in Australia, the USA and Europe. This has been evidenced by analysis of data from the International Earth Rotation Service (IERS) shown for a time period from 1992 to 2000 (Maxlow, 2005, p. 78).

Alternatively, we are able to estimate the extension of the Earth's crust nowadays going along with the increase in the ocean crust in the mid-oceanic ridges – which are about 70,000 km long. According to geodetic measurements the continents drift apart up to 20 cm per year. May we calculate only four centimeters as the average spreading rate of all mid-oceanic ridges. Then in one year about three square kilometres of new ocean crust is formed, representing a current Earth's diameter growth rate of 17 mm per year.

The breakup of the crust of our Pangaea-Earth happened at the beginning of the third phase of expansion, however, like a fast-break as it can be expected if the breaking point of the rocks is exceeded. The proof of a sudden crust-break is the simultaneous presence of various types of dinosaurs on almost every continent, which *should have been separated* by

Figure 63: Third phase of expansion. *The paleo-globes developed in 1933 by Christoph Ott Hilgenberg (TU Berlin) and re-constructed in 2001 from Professor Giancarlo Scalera at the National Institute of Geophysics and Vulcanology in Rome, Italy. The right is a glass globe with the present Earth and a smaller Earth in the interior, constructed by Klaus Vogel. The continents move away from each other, but remain fixed, despite the possibility of small shifts and / or twists as a result of increase in volume that is not exactly equally increasing.*

deep oceans long time periods before – if the plate tectonic time scale will be correct. One of several dinosaur examples is Majungatholus. These close relatives of Tyrannosaurus lived approximately 70 to 65 million years ago in Madagascar, which shall be an island at least since 150 million years. How a Tyrannosaurus could come across wide oceans from North America to the isle Madagascar? Such findings are piling up.

A giant frog called Beelzebufo was found in 2008 that had lived together with Majungatholus in Madagascar, but also in South America at the same time period. This discovery leads to a requirement that there should have been a land connection between Madagascar and South America at the end of the Cretaceous period, i.e. the end of the dinosaur era. (Evans et al., 2008). Constantly such controversial findings are increasing. Thus, the remains of a crocodile type have been discovered in Brazil, presumed to be 62 million years old, and it is said to have survived the extinction of the dinosaurs. The newly discovered fossils of *Guarinisuchus munizi* seem to be closely related to the crocodile, which once lived in Africa. The researchers say that this can be traced back to a transoceanic migration, because they had more primitive fossils of these crocodile in Africa, some younger ones in South America and further developed ones in North America (Barbos et al., 2008). Such findings of species on the "wrong" continent are showing that the plate tectonics timetable must be wrong. The splitting up of the

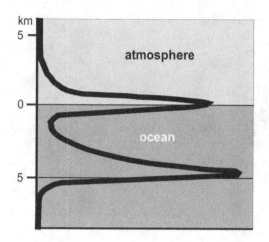

Figure 64: Equal ocean depths. *The two stages of the Earth's surface is a unique geophysical clarity: The Continental Shelf as the higher level appear at the boundary almost as a sharp tearing edge and protrude always in the same amount from the ocean floor. The continental shelfs can be joined together like a puzzle.*

The deep sea level as the other stage shows in all the oceans the same depth distribution. Only with Earth's expansion is it possible to explain this distribution satisfactorily.

continents should not have been imperceptibly slow as determined in the geological time scale, but should have occurred as a quasi-suddenly breakage. At the beginning of the dinosaur era, there were absolutely no oceans yet, and half of all the oceans originated after the extinction of the dinosaurs, according to the geological time scale.

At the time as the diameter of the Earth was just a little bit more than 60% of its diameter today, one might conclude that the gravitational force was correspondingly lower at that time, thus the existence of huge dinosaurs was possible at all. It is a fact that no major land animals can exist which is bigger than elephants, if we underlay present conditions. This is ignored because nobody can explain why up to 50 meters long dinosaurs could exist at all – if these animals have a body with muscles all over!

The long-necked sauropod dinosaurs could also not have been compact and muscled for various reasons, as was discussed in my book "Dinosaur Manual" (only published in German). The tiny head was much too small to let plenty of grass pass through the mouth, therefore this dinosaurs were not able to generate enough energy to move the muscles. Elephants which are busy for up to 18 hours with the daily food intake require at least 100 kilograms of plant food every day. These animals are the greatest possible form of life on land! Larger is not possible. A 30 or up to 50 meters long muscled dinosaur is not possible under today's conditions.

These sauropods used a trick. Instead of a compact body through and through there was a hollow section, because they probably had a very large "fermenting barrel" which functioned similarly on small-scale inside today's cows, said Professor Josef Reichholf (Zoological State Collection, Munich) in a personal conversation with me. This view was presented by me in a documentary of the scientific television program "World of Wonders" by "PRO7" first time on 22 September 2002: In the fermenting barrel, dinosaurs transformed the plant protein into bacterially protein and the animals were

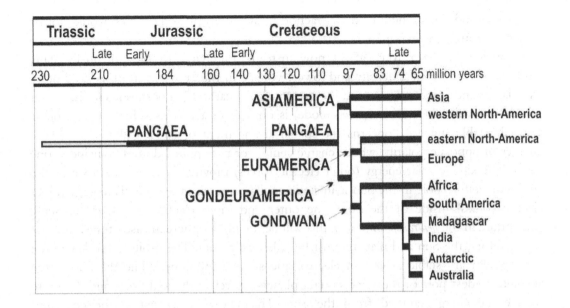

Figure 65: Dinosaur paleo-geography. *New discoveries of dinosaur fossils lead to a reduction in the time scale of plate tectonic processes, according to Professor Paul Sereno (1999). Because of new fossil discoveries since 1999, as stated in "The Dinosaur Manual" (Zillmer, 2002, p. 85), a further reduction in time can be concluded, which was then updated as follows in the book "The Human History Mistake" (Zillmer, 2010): New fossil findings traced back to a later break-up of the Pangaea (-Earth Figure 63). Black bars show the continued existence of current continents as it will be add up of new expertise's of the dinosaur paleo-geography: Pangaea had almost 100 millions of years longer existence than it is estimated by the geological time scale (grey bars). With "Gondeuramerica" a new preliminary continent must be defined; the northern continent Laurasia supposed till now must be deleted.*

therefore much lighter than is shown by today's reconstructions and needed *less* food. So we come to the conclusion that the strength of "gravity" must have not to be changed *drastically.* But how could take place a change of the gravity force?

Electrical Interaction

Instead of a mass attraction with compressive forces we had already alternative described an attraction by electrostatic forces in the solar system. In the electrostatic system, two bodies charged in the same direction repel each other, so that there are stable orbits. A variation of the stress state in our planetary system would change quickly the distances of the planets and the strength of the attraction (gravitation) pressing, which would not be possible through gravitation.

How does pressing force affect now instead of gravity? In the potentially state, electricity has its highest density and lowest voltage in the Sun – as in other cold poles like planets and comets. Therefore, the negative pole (cathode) is positioned in the center of the Sun while the positive pole (anode) is on the surface. This sphere as a whole is statically stable; since it concerns the spatial arrangement of two equally large electric charges of opposite polarity, whose charge centers or pole points do not coincide at one point. The kinetic solar energy (outer electric forces) moving from outside towards the cold temperature pole in the Sun is in static equilibrium with the potential energy (inner electric forces) inside of the Sun. The outer and inner electric forces are in static equilibrium in the *neutral spherical shell*, this will be the photosphere in case of the Sun.

Therefore, the Sun and planets must be spherically built. The orbits of bodies in the solar system are stable at certain electrostatic stress conditions. While the Sun is the absolute coldest pole of the solar system, planets are relatively cold poles and therefore receive the energy emitted from the Sun. This energy is stored in the core and materializes in the metallic hydrogen. These originated substances are streaming upward in direction to the *neutral spherical shell* with a continuous decrease in pressure and modification of substances during the entire pass. In other words, the inner forces are pressing the substances from the inside to the *neutral spherical shell* of the Earth and at the same time the external forces press centripetally from outside to the Earth i.e. into the *neutral spherical shell*. There the internal and external forces are statically neutralized.

In this "neutral" zone below the surface the Earth is at its warmest (see Figure 55, page 133), and the temperatures are decreasing toward the Earth's surface on one hand, and towards the centre of the Earth on the other hand. Therefore: Since the external force sizes are directed to the *neutral spherical shell*, we too, like dinosaurs earlier, are *pressed* against the Earth and not gravitational attract.

The pressure from the outside as a "gravitational effect" comes to an end in the *neutral spherical shell*. With a free path, a body would only "fall" (i.e. pressed) as far as into this sphere, but not to the center of the Earth. Conversely, a body from the Earth's core would be pressed in outward direction ultimately up to the *natural spherical shell*, regardless of the centrifugal forces caused by the rotation of the Earth.

Figure 66: Electrical monopoly. *The external and internal electrical forces are in equilibrium in the neutral spherical shell, as the negative pole is inside, inverse to the electron (Meyl, 1999, p. 101).*

150

The erroneous interpretation of the "gravity" is especially clear to see in space flights, since practically all calculations regarding a landing on other planets are obviously wrong, as almost all planned soft landing attempts of space probes fail or lead to obscure results. So, for example, the parachutes and braking rockets were miscalculated, and NASA inevitably changed to a rubber ball system for successful landings on the Mars. Even for the Moon, one required many attempts until enough empirical experience was collected. The first Moon rocket, the Russian *Luna 1*, should have struck the Moon, but missed it by 6000 kilometres, and the following U.S. *Pioneer 4* missed the Moon at a distance of 60,000 km, supposedly due to a defect. Including these two, in ten years only 21 out of 68 lunar missions were successful, until then in 1968 suddenly out of nowhere the first manned probe Apollo 8 circled the moon.

Even Mars, the space probes approached gradually. *Mariner 4* was the first space probe to reach Mars in 1964, it flew by at 9846 miles. 22 photographs were made from this distance. *Mariner 6* and *Mariner 7* decreased the distance in 1969 to approximately 3430 kilometres. After 14 attempts, in 1971, the probe *Soviet Mars 2* succeeded in swinging into the orbit, even before the American *Mariner 9* swang into the orbit on 14th November 1971 as the first artificial satellite. From 1960 until the end of 2005, 37 spacecraft's were sent to Mars, out of which only 13 missions were successful. The rest of the space probes were either partial success like *Fobos 2*, otherwise complete failures. Therefore, it was told of aliens on Mars, which catches or destroys spacecraft's . . .

An "gravitational effect" can be (roughly) calculated, but this empirical formula tells us nothing about the cause in itself! Formally the Coulomb's law clearly has *the same structure* as the law of gravity. *Both laws differ only in the principle of action*, on the one hand attraction and on the other hand repulsion which results in pressing onto the planets. This repulsion occurs between the planets of our solar system, if, under the laws the vortex kinematics, same type of polarity is facing to each other. If the planets and other celestial bodies are electrically conductive and have the same polarity, they cannot collide. On the contrary, they are maintained on certain predictable paths. So many small celestial bodies fly in "formation flight" with the Earth. Collisions of celestial bodies can therefore only take place if one of them is electrically neutral or the kinetic energy is greater than the electrostatic repulsion, but also if there have opposite polarities. However, it is also possible that could be changes in the orbits of the planets, especially in case the voltage condition and therefore the size of the solar *neutral spherical shell* changes through increased or decreased particle flow from the universe. There are reports of such "unstable" orbits from ancient cultures around the world (for more details see my book: "Mistake Earth Science", June 2007).

The electrostatic (Coulomb) interaction is not only responsible for stable orbits in the planetary system, but also for the cohesion of electrons and atomic nuclei in atoms and molecules and thus for all chemical and biological processes.

5 Chemical Energy and Life

The Earth was more than four billion years ago no hot magma ball. A research team led by Mark Harrison (2007) by the National Australian University in Canberra, examined the titanium content of zircons in igneous rocks from Australia and Tibet and came to the conclusion that the Earth at this time was a planet with relatively cool oceans and continents, and it was less hellish hot as previously thought, published in the journal "Geology" (vol. 35, July 2007, p. 635–638). Already during this first millions of years in the turbulent history of the Earth could have been existed life on our planet (Nemchin, 2008).

Partially Melted

Previously it was believed that the Earth was formed like a burning-hot body in an icy universe. In the meanwhile, prevailing scientific opinion is that the Earth, like the other inner planets, was gradually massed-together from solid components, which existed originally in a gas planetary disc. The "supposedly enhancing gravitational pressure" and the radioactivity should have the effect that the terrestrial matter gradually, but only partially heated above the melting point and was liquefied. The heating is then continued, partly by movement of the material released due to "gravitational energy" and by energy from spontaneous chemical reactions in the material mixtures, as well as the impact of meteorites.

Therefore, the geologist had no other choice to believe that petroleum was built much later out of biological remains, since hydrocarbons produced by high temperatures would have been destroyed in the Earth. But if these theories of a hot Earth are to be correct, then other planets like Mars would have been once hot, and there should be no hydrocarbons because there was and is no life on the surface. But methane was discovered also on Mars! May be in the future they discover ethane, in addition to methane which is a main component of petroleum. On Jupiter, Saturn and Neptune, as well as in the coma of Comet Hale-Bopp hydrocarbon was already detected. On the dwarf planet Pluto was discovered ethane ice and on Saturn's moon Titan is a lake to be filled with ethane, according to a report in the journal *Nature* (Raulin, 2008).

Hydrocarbons found on meteorites, as well as on Mars and other planets have been originated abiogenic because biological processes there have found no evidence so far. Mars can therefore – like the comets – not have been hot and liquid, but originally consisted of a low temperature condensate like the Earth and was also only partially melted. If there are methane-producing microbes in the interior of Mars, how scientists speculate due to lack of alternatives for the origin of methane, these are not able to produce such methane quantities of 36 kilograms per minute. In addition there are real methane clouds in the atmosphere (Mumma et al., 2009). Moreover it is not known that microbes can produce ethane in addition to methane!

The Earth should not take a planetary exceptional position. Therefore, on our planet existed hydrocarbons since the formation of our planet and these have been abiogenic originated. This view is supported by the results of seismic tomography, as shown by the three-dimensional images of the mantle. Hot magma is only to be found in certain areas and to depths of 400 km (Anderson / Dziewonski, 1988, p. 74 f.). But, should it not be hot throughout this sphere if a hotter Earth's core heats the mantle rock in deeper layers of the Earth's interior. That this view is wrong, showed the results of seismic tomography with three-dimensional images of the mantle because hot magma exists only in certain areas and occurs solely down to depths of about than 400 kilometres (Anderson, Dziewonski, 1988, p. 74 f.). Should it not to be hot also in deeper layers of the Earth's, if a *very hot* Earth's core heats the rocks of the mantle from below?

The model of the electrical Earth is consistent in line with the findings of seismic tomography, because the core of the Earth is cold and appears only in the *natural spherical shell* which is *partially* melted. Radiation exists therefore not in the core of the Earth, but in the *natural spherical shell* with magmatic rocks and is ultimately produced by cosmic energies. Therefore, it is technically possible to reduce this radioactivity.

With the patented *Patterson transmutation cell* (in particular U.S. Patent No. 5.672.259. 30 September 1997), a special electrolysis cell, is loaded radioactive nuclides. During the following electrolysis the measured decay activity takes place dramatically. On 28 May 1997 was published an experiment in the American television channel ABC, in which about 50% of an uranium-nitrate nuclear radiation was eliminated within one hour and another 13% more in half an hour (Gruber, 1998). In a few hours resulted in reductions of radioactivity of up to 80%. In these nuclear reactions could be observed *new* elements (in detail in: Meyl, 1999, p. 136 ff.).

This electrolysis cell can reduce the radioactivity of a radioactive material and simultaneously head is generated. The observed excess of thermal energy in comparison to the neutron flux may be explained to be a decoupling of "free energy" existing in space, according to some professionals (see Oesterle, 1996).

What is electrolysis? During electrolysis electrical energy will be converted into chemical energy. But this is actually the one process that takes place inside the cold Sun and the cold Earth – according to our previous discussion. Basically, the electrolysis is a reverse process of going on in a battery. For the case of splitting water into hydrogen and oxygen you have an eco-friendly engine for our cars. However the problem is, that it is far cheaper to produce hydrogen from petroleum or natural gas. No one wants an eco-friendly machine producing energy costs, which can go nearly to zero.

It is interesting also to note that in the electrolysis cell new elements emerged. Consider an interesting example, namely the production of the shell of a hen's egg. This consists of calcium, and as long as chickens eat calcium containing food, one sees no problem. Now the chicken feed was deprived of substantially all calcium. But still the hens produced eggs. Then, other substances were deprived from food. It eventually found that only at the removal of silicon, the production of eggs stopped. It was concluded that the chickens produce calcium itself from silicon, probably by using carbon. This example is often cited for an implementing process of *cold fusion*, also called *low-energy nuclear reaction* (LENR), refers to the hypothesis that nuclear fusion might explain the results of a group of experiments conducted at ordinary temperatures (room temperature).

Cold fusion, you have to note, is officially denied, and certain economic interests play a role. Cold fusion was announced by the two chemists, Dr. Martin Fleischmann and Dr. Stanley Pons. They have proved with experiments in 1984 at the University of Utah for the dissociation of heavy hydrogen (deuterium) by applying a voltage (electrolysis). In these experiments, arose more energy than it was consumed, and excess heat created (Fleischmann et al., 1989). Although some laboratories in Europe, the USA and the USSR reported to have reproduced the Fleischmann / Pons effect the scientific and public opinion struck immediately into denial. In November 1989 a commission of the *United States Department of Energy* (DOE) came to the conclusion that the current evidence for the discovery of cold fusion as a new *nuclear reaction* is not convincing (Maddox, 1989).

Inside of the Earth's core, however, could be performing cold fusion converting energy into matter. "In the outer core are formed different atoms and isotopes, emitting energy more than absorbed while the fusion process is going on" (Meyl, 1999, p. 19). Therefore in the Earth's core arises warmness. This temperature is required to maintain a maximum conversion chance and the stability of open chemical systems around 310 Kelvin or 37 degrees Celsius Celsius. This is the temperature of a human body and therefore no fortuity. The statistical chemistry offers the possibility to simple describe the geochemical systems as Otto Oesterle has done in detail (1997, p. 90 f.). With the increase of pressure with depth in the Earth, increases the optimal temperature of transformation processes and chemical self-organization according to a nonlinear dependence (see Fig. 67).

Mainly inside the Earth is going on the forming of simple hydrogen atoms, which combine with carbon and oxygen atoms on their way out directing to Earth's surface. That process will lead to a chemically forming of perhaps one cubic kilometer of water, which is new added in the Earth's interior every year (Fig. 67). Water plays a major role in

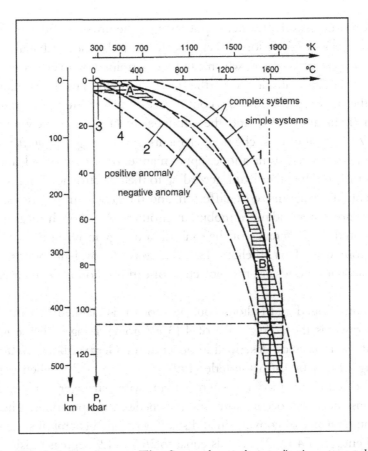

Figure 67: Chemical Development. The figure shows the qualitative suspected dependence of pressure (P) and depth (H) for the optimum temperature of self-organization of chemical systems in the depths of the Earth (1) and the temperature inside the Earth (2), for the life of the Earth's surface (3) and on the floor of the Atlantic Ocean (4). With increase of pressure in the depth the optimum temperature of chemical transformation (36.5 degrees Celsius) and chemical self-organisation is rising according to a nonlinear dependence (line 1). The dots number 3 and 4 are corresponding to the life on the Earth's surface and on the bottom of the ocean in the "black smokers". Line 2 shows the dependence of the average temperature of the Earth's crust from the depth. In the special zones A and B, there where cross over lines 1 and 2, and in this areas with positive temperature anomalies are existing optimum conditions for the chemical self-organisation, completion and arrangement of materials. In zone A, there are the biosphere and the deposits, so the probability from its formation decreases in principle with the depth. Zone B is so far very weak investigated, but it may be connected with the emergence of biogenic hydrocarbons. The hydrogen atoms formed inside the Earth's connect on their way up with carbon and oxygen atoms, will lead to a forming of "new" (juveniles) water and primary materials for oil and natural gas, which will be aggregate and then accumulate in "traps" and build deposits near the surface or sputter out of the Earth or the seabed. From: Oesterle, 1997, 91 ff. (at Technical University Berlin).

all geochemical processes. Hot water vapor changes the physical and chemical properties of the adjoining rocks (alteration). There are formed also new minerals with a higher content of hydrogen and oxygen, which can be determined by microscopic observations of the rocks. In addition, inside the Earth affects electricity. The distribution of chemical elements in the upper mantle and the continental crust indicate the thermoelectric field of the Earth (Lehmann, 1994) and the continual emergence of new atoms. In other words, *without* electrical effects, chemical elements are *not going to be originated ever-new* and have also not accumulated at the anode (upper mantle) on the one hand, and the cathode (continental crust) on the other hand, according to their electrochemical series.

The elements normally move by diffusion, and the geochemistry has determined that some elements are faster and more mobile than others. As a rough measure of mobility can be considered the difference of the ionization energy in relation to the size of 8.26 eV (= electron volt). Thus, helium has the *weakest* covalent bonds, but is very mobile. The strongest bonds are the molecules of carbon monoxide; their energies are at 10 eV.

To avoid complicated and tedious considerations it is claimed that the atomic bond energy tends towards the average size of 4.13 eV going along with the growth of the complexity of the molecules; described in detail in the German book *Goldene Mitte: unser einziger Ausweg* written by Otto Oesterle (1997, p. 53 ff.): "But when covalent bonds occur more often with medium energy than with stronger and weaker energy, transformations must also occur more often at medium temperatures. The temperature of the maximum chance of conversion and stability of open chemical systems (thus with a mean bond energy of 4.13 eV) . . . is equal to 36.5 ± 1.9 degrees Celsius" (ibid., p. 90 f.).

Overall, we identify a form of system development, where shaping and formative influences of the elements of the system emanate from themselves. With this process of self-organisation certain closed systems are developing stable forms of behaviour and conditions. Such behaviour of a closed system that acts independently, acts on itself and becomes a starting point for further developments.

Therefore, the body temperature of humans is purely coincidental to nearly 37 degrees Celsius. This temperature corresponds to the dynamic balance of energy-consuming and absorbing processes. The body temperature is thus neither a product of chance nor the result of an evolutionary development, but a coerciveness of stable systems.

As noted earlier, this is the maximum temperature conversion probability "only for" normal pressure. With higher pressure-levels which increase with increasing depth, the optimal temperature of the transformation and chemical self-organization expends according to a nonlinear dependence. In this case the atomic bonds of petroleum, for example, have not the average bond energy of about 4.13 eV, but 5.5 eV. The result is an optimum petroleum temperature of 80 to 90 degrees Celsius. Taking into account the

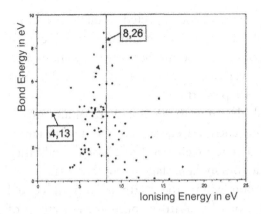

Figure 68: Bond energy. *The strongest atomic bonds form atoms with an ionization of 8.26 eV. In this case these are not only substances with the highest bond energies formed but also with the highest melting temperatures and the largest thermodynamic stability systems.*
According to Oesterle, 1997.

Particular pressure at a certain depth, the accumulation of petroleum takes place as a result of an inorganic (mineral) process going a "systematic" way – called the *abiogenic petroleum (hydrocarbon) origin*. In this way hydrocarbons are produced at a depth of 200 kilometers in an area that is marked B in Figure 67.

At Wikipedia encyclopedia you can read: The biogenic theory for petroleum was first proposed by the Georg Agricola in the 16th century in Germany, known as "the father of mineralogy". Various abiogenic hypotheses were proposed in the 19th century, most notably by Alexander von Humboldt, the Russian chemist Dmitri Mendeleev and the French chemist Marcellin Berthelot. Abiogenic hypotheses were revived in the last half of the 20th century by Russian and Ukrainian scientists, who had little influence outside the Soviet Union because most of their research was published in their native languages. The theory was re-defined and made popular in the West by Thomas Gold, who published all his research in English.

This view of an abiogenic origin of hydrocarbons will be only correct if the Earth was and is not melted through and through in deeper layers. This means that there exists only melting nests or in other words the Earth is melted only partially, because otherwise the volatile components would have risen to the surface and disappeared over all other materials. If the Earth was hot in the depth over all, it would have been no reason to expect a further influx of such elements from the deep. Just the reverse is true for an initially cold instead of hot body. If the Earth is cold in the depth then a gradually warm–up of layer by layer will be persisting and the volatile substances were driven off very slowly: *The Earth carries a process of outgassing.*

At the opposite view, whichever is taking in account an initially hot Earth, most of the fluids would have been brought very early to a low energy state with the consequence that later on it couldn't come to chemical reactions with a corresponding energy release.

If the Earth was and is not hot but relatively cold, the expected result is in fact the today observed picture of the composition of the upper mantle and the lithosphere: If the volatile constituents escape, they would often not be in chemical equilibrium with their environment and may therefore offer even today a source of chemical energy (Thomas Gold in "Annual Review of Energy", 10, p. 53 ff.).

If we take in account a hot instead of could Earth in the depths there cannot exist a chemical energy at all. In this case there is no chemical energy left and all substances within the Earth have come already towards a state of chemical equilibrium and "then the only energy source for Earth life would be sunlight falling on the surface. An understanding of the oxidation state of carbon within the Earth is thus of central importance. The question of the stability of the Earth's primordial supply of hydrocarbons against oxidation – that is, against combining with oxygen contained in the silicate and other minerals of rocks – is intimately connected with the details of the outgassing process. If the gases ascend in regions of magma, then (as we have already discussed) chemical equilibrium between the hydrocarbons and the magma would be approached, and this would usually favor oxidation of the hydrocarbon gases. Thus is the surprise that volcanoes generally emit carbon mainly in the form of CO_2, with only minor amounts of methane, CH_4", as Thomas Gold (1999, p. 48) correctly stated.

In contrast, the modern geology of 19th Century until the second half of the 20th Century was influenced by the idea that the Earth was once an initially melted celestial body of blistering heat. Therefore, the geological survey of this view is fundamentally shaped. Although today has become clear that the material of the Earth was only partially melted, the geological hypotheses have not previously been thoroughly revised accordingly. But this would be needed stringently after such a serious change in the basic assumption. "Nowhere is this more evident than in the discussion of the origin of volatile substances on the surface: the water of the oceans, the nitrogen of the atmosphere, and the carbon-bearing fluids that appear to have been responsible for a great enrichment of the surface with carbon" (ibid., p. 48).

Previously it was widely believed that hydrocarbon molecules like methane decay at temperatures higher than 600 degrees Celsius and 300 degrees Celsius are enough to decompose the heavy hydrocarbons in place. If the Earth was actually melted earlier, so the origin of the today present hydrocarbons could not have existed in very great depths. If this is true there is no alternative to the theory of a biogenetic origin of petroleum and natural gas, why we are convinced of their fossil origin and therefore limited supplies.

But if we take into account a cold instead of hot Earth we get exactly the opposite statement of the conventional theory: the seemingly unthinkable abiogenic origin of hydrocarbons. The largest part of the alleged "fossil" fuels are originated abiogenic (inorganic) but not biogenetically! Extensive spectral analysis has been proven that carbon is the fourth most abundant element after hydrogen, helium and oxygen which

Figure 69: Iapetus.

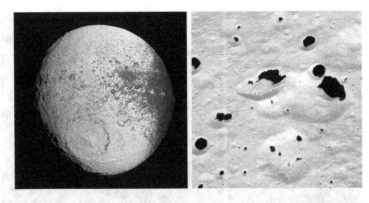

This Saturn's moon has an albedo from 0.03 to 0.5, the greatest brightness contrast of all bodies in the solar system and has large areas, white as snow, and some of which are partly black as ebony. There were smaller bright impact crater observed on Iapetus with a diameter of 30 to 60 m and a depth of about 10 meters, "the apparent upper crust of dark and bright material penetrate from the underground ... The black material appears only a thin crust to be formed on the Moon "(Kehse, September 18, 2007). The crater should be in pockmarks, from which gases break out. The black spots are, in this case, from black carbon, which is particularly produced by the incomplete burn-up of hydrocarbons such as methane. Images: Cassini spacecraft (NASA).

are occurring in the universe. But carbon occurs on planetary celestial bodies usually in combination with hydrogen, thus as a hydrocarbon in gaseous, liquid or solid form. It was proven that methane and ethane occur on other planets and their satellites in our solar system – up to this time this occurrences were thought to be impossible. Not only the comet Halley is covered with carbon, so its surface appears as covered with pitch black. Do these carbons originate from the inside of the celestial bodies?

Surprising Explosion

Entirely new information about comets has delivered the three space missions, *Deep Impact, Deep Space 1* and *Stardust* since 1999. But instead of finally revealing the true nature of comets, the sometimes conflicting data from these missions have put almost everything into question what scientists thought to know about these fascinating objects.

From the view of astronomers the detail photographs of the comet *Tempel 1* exhibits a shocking appearing diversity of its surface, such as smooth surfaces alongside rough and various craters.

Since the measure of the albedo was determined to be 0.04, the surface of *Tempel 1* is absolutely black such as the surface of the comet Halley. *Tempel 1* also has a high content of hydrocarbons. On 4 July 2005 the nucleus of this comet was struck by a 372 kg heavy, refrigerator-sized impactor of copper, which was brought by the Deep Impact

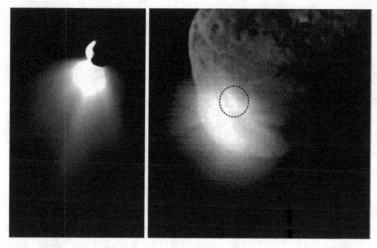

Figure 70: Temple 1. *The* left picture *was taken 45 minutes after the impact of the impactor. The comet is still shining out of his scraped wound.* Right picture: *Approximately five seconds after the impact on Tempel 1 started a second area to radiate (dashed circle).*

spacecraft. Shortly after the collision between the impactor and *Tempel 1* a brief but very bright flash could be observed in which the missile was destroyed in an explosion. The brightness analysis shows that the comet first lit up intensive in the minute after the impact and in the next six minutes the brightness still increased. But then, for the next 10 to 15 minutes the comet gets even brighter, and only after 45 minutes the brightness was slow-going down. A long-term effect had had the impact obviously not, because on 9th July 2005 *Tempel 1* behaved as usual before the impact. The probe's spectrometer instrument detected dust particles like carbonates, complex organic compounds, silicates (e.g. olivine) and clay minerals. In addition to the cloud of dust developed some different gases. There seems to be more volatile than solids, so comets should be named icy dirt balls rather than dirty snowballs. In addition to water and carbon dioxide detected at the time of the impact the probe detected water, carbon dioxide, methanol, hydrogen cyanide (hydrogen cyanide) and ethane. These chemicals sprayed into the interplanetary space, where the proportion of ethane even increased after the explosion. In this cloud then was detected amongst others ethine (acetylene) and methane. Later on, there was emitted only water and carbon monoxide, perhaps an indication of an incomplete flash burn of hydrocarbons. Thereby was produced water and carbon monoxide, or carbon (carbon black), which is responsible for deep black colour of comet's nucleus.

X-rays were not measured, but ultraviolet light, according to the interpretation of the researcher originates from so-called hydroxyl radicals, a decay product of water. Within half an hour after the impact, the brightness of the hydroxyl (alcohols and phenols) had quintupled, before the issue went back in the ultraviolet range.

With the "dirty snowball" theory cannot explain the comet phenomena obviously. At the surface of *Tempel 1* were measured temperature of −13 and +56 degrees Celsius, much warmer than the surrounding interplanetary space. The warmer areas were in the range of the most favourable temperature for chemical transformations; while in colder places is it possible that originate water ice on the surface. And for the *first time ever*, on the surface of a comet was actually discovered water ice, even though only in small areas. The question is why such a dirty snowball can become so warm in the ice cold space without a development of brightness? Is it actually a snow or an ice ball? Analogous to the Sun a comet should have a *neutral spherical shell* just below the comet surface. Therefore the sun-facing side of the comet nucleus is naturally warmer, because there the kinetic electricity (solar wind plasma) coming from the Sun is acting strongly.

As observed for Sun and Earth, it must be colder inside these conductive celestial bodies with increasing depth under the neutral spherical shell. This is proven by the large amounts of water vapour, which were registered after the impact: This is an evidence of vaporized volatile components, which can only exist in a *cold but not a hot* environment.

Comets are mostly made of stony materials contrary to the hitherto existing scientifically view. This new view is confirmed by the American astronomer, Professor William Reach of the *Spitzer Science Centre* in Pasadena after an examination of 29 comets. Some stone-existing asteroids might therefore be ex-comets which have lost its electrical conductivity. Therefore, it is not surprising that the comet *Wild 2* has a similar composition like a typical asteroid, proved by particles which were getting out of this comet tail and brought to Earth by the Stardust spacecraft in 2006 (Ishii et al., 2008).

Due to the emitted substances it can be concluded that in the interior of *Tempel 1* carbon respectively carbon monoxide must be present. The pitch-black surface of the comet is covered as a result of incomplete combustion (oxidation) of hydrocarbons which *Tempel 1* emits plentiful. Therefore those celestial bodies can appear totally or only partially black and these in areas, exactly where gas jets breaks out.

The NASA probe *Deep Space 1* took photographs of the only eight kilometers long comet *Borrelly* in 2008. This celestial body had even just an albedo of 0.03, and in some cases of only 0.007. These areas are so pitch black, that almost no light is reflected. As previously outlined the geochemical processes and the existence of carbon or soot are not officially recognized, scientists believe alternatively that on the comet should be exists a so far *unknown* mineral with a reflectance value (albedo) near zero. *With a wrong theory you can generate more mysteries than it solves!*

On 2 January 2004 attended the NASA spacecraft *Stardust* the short-period comet *Wild 2*, which shall be 4.5 billion years old. The close-ups show a comet nucleus with about five kilometers in diameter. The surface has an albedo of 0.04 and is as black as the comet *Tempel 1*. It's rough surface is covered with boiler-like hollows, but the edges are steep and rugged. Relatively to the total size of the comet can be identified relatively large besides small structures. It is believed that these are meteor craters – the usual scientifically explanation. But during the flyby, the spacecraft detected at least ten active, directed gas streams. Therefore the alleged "meteor" craters may have been formed alternatively as a result of gas outbursts, which would explain the steepness of the crater walls. If the pockmarked surface of some or all celestial body is formed mainly due to gas outbursts, then the age estimation of such objects is wrong, because the number of craters per unit area represents a measure of the age.

In any case, the previously described impact gas cloud on *Tempel 1* surprised above all by a completely inexplicable high content of organic, mostly unidentified substances. They appeared mainly in the first phase of the outbreak. Moreover, it was measured too much water vapour as such could have arisen going along with a purely direct transition from solid to gaseous state (Mumma et al., 2005).

What exactly happened at the chemical explosion on the comet *Tempel 1*, cannot currently be completely clarified, because the researchers were not prepared for such a wide event. According to the current opinion, it was simple just a snowball. But it is clear that the comet *Tempel 1* consists of a large part of hydrogen, oxygen and carbon. There were detected numerous organic compounds which solely can be built up of carbon and hydrogen. Even *before* the collision the researchers found the (aliphatic) hydrocarbon ethane, a major component of natural gas after methane.

In addition to water was methanol present (Mumma et al., 2005), which can be built up of hydrogen and carbon dioxide or carbon monoxide (with simultaneous formation of water). In addition to this, a few minutes after the explosion it was discovered ethyne, also known as acetylene, which consists of carbon and hydrogen. This colorless gas has also been detected in the atmosphere of Jupiter and in interstellar matter also. At high temperatures of 2000 degrees Celsius acetylene can originate from methane, if oxygen is *absence* in a gas discharge. In larger quantities than acetylene, methane was present in the gas cloud, before it was detected mainly just water and carbon monoxide. In deeper and therefore cold layers of the comet methane is present in liquid form, because it melts if a temperature of −182.6 degrees Celsius is reached at atmospheric pressure. On the surface of *Tempel 1*, it is warmer, and therefore methane occurs there in gaseous form. As long as sufficient oxygen is present, methane will be oxidized respectively burns up completely, and with this process is going along a development of carbon dioxide and water. If the oxygen supply is reduced, then may occurred carbon monoxide and water. Exactly these substances have been demonstrated in the gas cloud at last of the impact event. Or alternatively, there are occurring water and carbon, respectively soot, so *Tempel 1* also has a black surface. Inside the comet, the abundant hydrogen reacts with carbon monoxide, which has a very strong atomic bond. Thus methane and water come into being, whereby we are able to explain the origin of rich water supply not only on the comet *Tempel 1* . . .

The sequence of chemical reactions is still unclear, since on the one hand contained the dust-gas cloud unexpectedly large amounts of organic substances and on the other hand electrical energy is officially disregarded. In addition, the pressure and temperature conditions are poorly known. At the beginning of the explosion the disgorged materials are to be a temperature of more than 700 degrees Celsius. Some authorities detected up to 3500 degrees Celsius for the ejected molten material in the surge.

As in the gas cloud water was continuously present, the question arises as to what physical state it was? Observations with a telescope of the NASA satellite SWAS (*Submillimeter Wave Astronomy Satellite*) show surprisingly little water vapor in the released material. The researchers believe that it was rather "damp dust" than the beforehand supposed "dirty snow". Water in liquid form on comets was not previously suspected.

A breakout of light intensity can arise due to the sudden triggering of volatile components, since the ignition temperature is lower for this than the temperature measured in the explosion cloud. Delayed by heat conduction, the critical temperature can be reached later on, which can trigger subsequent outbursts. These substances are unevenly distributed below the comet's surface, because on the nucleus surface of bright comets are only small areas highlighted active. Thus also hydrocyanic acid (hydrogen cyanide) was discovered and it is assumed that the dust might contain acetonitrile (C_2H_3N). Therefore, it is thinkable that together with oxygen, explosive mixtures can be formed and ignited by electrical potentials.

Explosive Outburst

If comets pose dirty snowballs, then there is no reasonable explanation for explosive outbursts that occur after the comet has left the Sun's closest point and is already on the way back to the border areas of our planetary system for some months. Exactly this unexpected event happened in October 2007.

The comet *17P/Holmes*, which was named after its discoverer Edwin Holmes already 6 November 1892, reached the nearest point to the Sun which lies still outside the orbit of planet Mars. On 14 May 2007 this comet turned back in direction of dark outer planetary regions after developing his normal brightness due to its distance to the Sun, as usual, and this very inconspicuous. Only with powerful telescopes, *17P/Holmes* was able to catch as a faintly spot in far distance. This comet's passage was therefore already forgotten, as *17P/Holmes* four months after reaching the point nearest to the Sun, began to shine brightly on 24 October 2007.

About 48 hours after starting the explosion of brightness, this comet was almost one million times brighter than before. The comet appeared peculiar star-like, with *no* apparent tail at all. Within the next two days the coma dramatically increased at the same brightness. With an expansion velocity of about 2000 kilometers per hour, the diameter of the inner coma was brought up to a diameter of about 850 000 kilometers, while that of the outer coma was actually about two million kilometers.

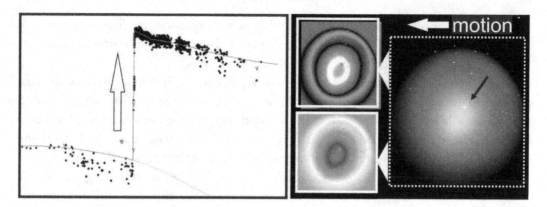

Figure 71: 17P/Holmes. *The graph (left) shows the changing brightness from May 2007 to May 2008. On 2 November 2007 taken photo (right) has been electronically processed differently by author (= two inserts), creating a multiple-ring structure is visible. The black arrow points to the core, which has a diameter of just 3.4 kilometers.*

A few days later increased the diameter of the inner coma of even a million kilometers (= more than 20 arc minutes). The tail was quite short, and on 8 November 2007 there was still a spectacular breakaway of the comet tail, and some days later he was completely gone. The diameter of the inner coma grew even more strongly up to 30 arc minutes middle of November. The peculiar yellow color had now given way to a blue hue. In December, the ghost was gone.

17P/Holmes was visible in 1892 with the naked eye when the British astronomer Edwin Holmes discovered this comet. Also at this time occurs an outburst with following fading. At that time, however, occurred a second, weaker outburst a few months later. In the years 1899 and 1906 again, *17P/Holmes* came near the Sun. However, he remained relatively faint throughout his entire approach to the Sun. Then this comet was considered lost. It was not until 60 years later, he was rediscovered.

Why is the detailed description of these events? The conventional comet theory has absolutely no solution to the riddle of brightness outbursts in the years 1892 and 2007. Also the in 1858 discovered comet *41P/Tuttle-Giacobini-Kresák* radiated several times brighter than expected at one of its approaches in 1973. As in the magazine *Sky & Telecope* discussed in detail on 15 November 2007, satisfies none of the offered explanations of experts for the outburst of *17P/Holmes*. Believed was a collision with a celestial body. Statistically, such a collision is very rare, but three one after another are impossible to explain the multiple outbursts.

Also a breakup of this comet can be no satisfactory explanation. Such an event could be observed as early as 1995, when the comet *73P/Schwassmann-Wachmann 3* broke into several pieces. The comet surprised the astronomers with a sharp, unexpected rise in

Figure 72: 73P/Schwassmann-3.
Several times, the comet broke apart.
1995 and 2006, the breakup of the
comet parts brought force in up to a 250-
fold increase in brightness – also the
fragments shone bright.

brightness up to 250-times in relation to its normal brightness.

The unusual thing about the sudden outburst of *17P/Holmes* is – next to the explosive behavior – the spherical shape of the light phenomenon, especially since this comet with other approaches always trained a typical comet tail. Principally, something must have been different in 2007, but also in 1892.

The solution to the riddle is sure not to search in a purely classical mechanical or physical-chemical scenario, but rather electric and magneto-spherical phenomena's play an important role. Unfortunately we know neither the electric nor the magnetic conditions at the time of the unusual brightness appearance. It could be either a magnetic reconnection process, in which the structure of the magnetic field – for example during magneto-spherical storms – abruptly changes and large amounts of energy are released. A reconnection can occur when a strong, variable magnetic field is present in plasma. It is a process of forming a potential structure, whereby magnetic energy will be transformed into a kinetic energy and by what is caused an increase in speed of the particles (electrons) in the space of plasma.

Magnetic reconnection in the solar corona will be responsible for solar flares. In the Earth's magnetic field, this contributes to the appearance of the aurora borealis. Without further addressing the phenomenon of the aurora, perhaps a special form is interesting, which is called corona. This is an aurora with radial rays in the magnetic zenith, which run together in a star-like point.

Without having now found a satisfactory explanation for the unexpected flash of comet *17P/Holmes* will be nevertheless clear that there is not a thawing snow ball, but discharge processes.

So far, 140 different molecules have been identified in space, in interstellar clouds and forming regions of stars. A large part of these molecules are organic, that is based on carbon, the most important element of the terrestrial biosphere. Carbon compounds form the molecular basis of all life.

Beginning of 2008 amino acetonitrile has been detected within *Sagittarius B2*, a forming region of stars. It is a chemical relative and possibly direct precursor of the amino acid glycine, also called amino acetic acid (Belloche et al., 2008). It is the smallest and simplest amino acid, an important component of almost all proteins and nodal point in the metabolism of living organisms.

We have found that many organic molecules are formed on comets, which are considered to be bases modules of life. These occurred due to various chemical processes going along with gas eruptions and electrical discharges. In the early 1950s, Stanley L. Miller undertook some experiments in this way. He sent electrical discharges through a similar mixture of terrestrial primitive atmosphere, which consisted of water, ammonia, methane, hydrogen and carbon monoxide. After several weeks could be detected amino acids and other substances, the basis modules of biomolecules. According to current opinion, however, the original atmosphere had of a different composition.

But Miller's experiments for the production of amino acids may not be representative because it was done in closed (!) vessels. Oceans and lakes form homogeneous media in which ingredients are evenly distributed in chemical equilibrium with each other. Under this assumption, the probability is nearly equal zero that simple organic substances were randomly organized to build a busy cell. In such a system does not develop a systematic arrangement (order), but rather disorder, if not a stability system prevails – according to the laws of thermodynamic.

Only under special physical system conditions is it possible that order increase respectively entropy decrease. One of those conditions is that the molecules are not distributed across the entire system and also may not be in chemical equilibrium, since otherwise lack the driving force to enter into reactions. Chemical substances have to be present in large concentration differences. "This is hardly compatible with the postulations of non-equilibrium thermodynamics and therefore behind all models of a solely terrestrial origin of life is to set a big question mark. The disposal of biological precursor molecules from space *changes nothing* in principle to the situation, at least as long as you accept, that this molecules would dissolve into the primordial-waters and only the pool of biological base materials increased "(Kissel / Krueger, 2000, p. 66).

On the surface of our Earth or seas biological life cannot have arisen by random reactions, which generated certain components in appropriate concentrations: "Thus was the origin of life on Earth not only an extremely improbable freak of nature, which needed hundreds of millions of years to give this chance a shot – according to the principle of trial and error" (ibid., p. 71). Therefore is recently re-discussed the panspermia theory, which already advocated in 1906 by Nobel Laureate Svante Arrhenius and lastly represented by Fred Hoyle. But the idea that simple biomolecules are cosmic indicator germs which have provoke the origins of life on Earth in the form of a starter kit – after the contact with liquid water – could never become accepted: Rightly, from the reasons we have already discussed.

Only by the findings presented here, which were obtained in the study of comets, it begins to show that on the surfaces of comet nucleuses perhaps arise amino acids. Although if this assumption should not to be confirmed, Kissel and Krueger (2000, p. 64–71) describe correct that the cosmic dust particles contain precursor molecules of all classes of substances, which are of importance for the biochemistry of living organisms. All components of the genetic molecules are arranged in dust particles of comets!

If it is not possible to detect amino acids on comets, then can help substitutable the proof of so-called unsaturated »nitriles occurring in the interstellar gas. Their reaction with ammonia – an omnipresent cosmic molecule – leads to amino nitriles. Of these, however, is known to form spontaneously amino acids on the so-called *Strecker synthesis*, if liquid water is present. The nitriles itself are able to react with water to build fatty acids. Consequently, they constitute base materials of a large family, the source material for make-up of living organisms: the fats (lipids) "(ibid., p. 68).

Apparently all the important biological molecules for life are contained in the dust of comets. However, this is not sufficient to develop life, although it seems that prevail ideal temperatures around 37°C in some areas on the surface of comet *Tempel 1*. At this temperature the chemical world achieved the highest complexity by combining different types of atoms. Water is the sole solvent which is liquid at this temperature and of all liquids dissolves well the largest number of substances. In addition, carbon is abundant in comets, which may form at 37°C long chains of its atoms. Carbon is included in all earthly creatures: All living tissue is composed of (organic) carbon compounds.

Therefore is, as already mentioned, the body temperature of humans not by chance around 37 degrees Celsius. At lower temperatures the metabolism is too slow, resulting in higher number of copy error, and living things age faster. 37 degrees Celsius connoted a maximum stability and durability. "All very complicated chemical systems in our universe can only exist at this temperature," provided that there a pressure of one bar prevail (Oesterle, 1997, p. 122 ff.).

The tolerance range of the temperature of living organisms depends on the complexity. The more complex the system, the more intense will be the metabolism, which apparently finds the best conditions at 37 degree Celsius (310 Kelvin). Compliance with this temperature is best realized in warm-blooded animals. The more complex an organism is the more sensitive he behaves with respect to temperature fluctuations. Therefore, the temperature variation is possible in a healthy human body, only about one degree to the optimum temperature, while the temperature range in bacteria increases to ± 20 degree Celsius.

Accordingly, it may be calculated from the temperature limits of a system, how big its complexity. In case of minerals this complexity is 1, 20 to 30 at bacteria and about 300 at humans (Oesterle, 1997, p. 123). If you consider that minerals can arise spontaneously, we must ask, why not also more complex chemical, perhaps also biological systems should arise *spontaneously*?

The approach of Kissel and Krueger (2000, p. 69) is based on a calculation of NASA, that the ancient lakes on Earth reached about 0.1 percent of the comet materials, where the porous grain structure was retained. By penetration of water into the loose particle structure now could be has started the prebiotic chemistry. But there still remains the objection that there was the water of lakes and seas in chemical equilibrium. Therefore, the formation of non-equilibrium thermodynamics was very unlikely, almost like winning the lottery with three super numbers.

If we take into account the theory presented by Kissel and Krueger, the jump start of biogenesis at least could had an off-chance. These authors share the opinion brought forward in my books: "None of the currently over one hundred biogenesis models allow nature to perform such a large number of experiments to enforce the origin of life at the stage of complete proto-cellular systems in such a short time period."

Figure 73: Geological self-organization. Stratification structures in geology – as visible in the sandstone samples – are based in many cases analogous to periodic structures in the physical chemistry on internal processes of self-organization. The classic example is the 1896-discovery of Liesegang rings, caused by a complex interplay of diffusion, chemical reaction and precipitation processes. In the left image formed by analogy with Liesegang rings concentric bandings to palm roots in Münzenberger sandstone (Germany). The right image shows a pattern of "chemical action at a distance" in the same rock plate. There is a strong argument for the adoption of a diffusely-related genesis (origin) of a precipitation pattern in a lemniscate-form (a horizontal figure 8) between two adjacent root channels. After Liesegang, R.E, in "Umschau" (Germany), vol. 34, 6/1930, pp. 102–103, quoted in: Krug / Kruhl, 2001, p. 290 f.

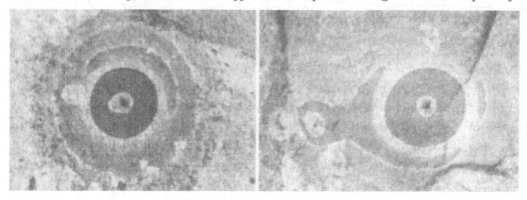

If one now conclude that the abiogenesis had started from the dust of comets may represent the only viable solution to this problem, then this should be correct if we stringently underlay the conventional world view, particularly a hot Earth. But instead of this, if we assume a cold Earth, is there an alternative possibility for the origin of life?

In the previously discussed spontaneous generation of life, which came from outer space, important components are included, which we have to take into account with the following alternative approach. At first there is self-organization as a principle that is often not fully appreciated in its scope. The precondition will be a chemical imbalance in the "primordial soup", which is an open system. In this case, the influences of shaping are starting from the elements of the organizing system *itself*. In other words, far from a chemical equilibrium on its own are occurring new stable, highly efficient structures and sequences. This will create an order out of chaos, without a vision is present from the entire development. There are very good examples in many knowledge fields, not only in physics, chemistry or geology (see Figure 53 and 54, p. 128 f.).

In a chemical imbalance in the primordial soup, there are certain stationary concentration gradients, especially a concentration gradient between two points. Thus the diffusion of materials occurs due to a concentration gradient. For our purposes also an electrochemical gradient must be considered, which a concentration gradient of dissolved ions constitute. According to our previous observations, electrochemical gradients are present in comets, where it is possible to originate new substances, respectively aromatic and organic compounds along with electrical discharges.

Consider a simple example: bacteria are not able to choose a direction to move always straight ahead. A bacterium moves in alternating tumbling and swimming movements. Within a concentration gradient bacteria adapt to this environment and are following a direction, either toward the attractant or away from the pollutant. In this case, it swims in a straight direction longer before tumbling occurs. If it goes in the wrong direction, it tumbles rather quickly and proposes a different direction. So bacteria are able to reach the range of higher concentration of the attractant very effective. The choice between swimming and tumbling is equal to the principle of a computer, because a bit as an information unit has the value 0 or 1, which adds up to either tumbling or swimming in our case. In higher organisms, which have a brain, the chemical reactions of the bacteria appear in the form of forgetting, the starting of movement and a quorum (Berg, 2003)

All what has been discussed has nothing to do with life, but it is a "chemical evolution" mentioned process. This, however, can only develop where predominate moderate temperatures are ruling, whether in space or on Earth. As we already had recognized, the optimal temperature is nearly 37 degree Celsius.

Not only on comets apparently are ruling optimum temperatures, but also on the Earth, if we do not believe that the early Earth was hot, but alternatively consisted of a low-temperature condensate like comets represent. Optimal temperatures for life – when

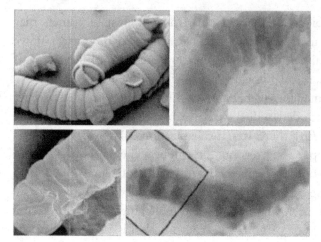

Figure 74: Structures. *Comparison of synthetic inorganic silica-carbonate structures (left) with 3.5 billion years old microfossils from the Warrawoona Formation in Australia (right). According to García-Ruiz et al., 2003.*

viewed with the geological time table – seem to have ruled promptly after formation of the Earth. Metabolic footprints in metamorphic *Isua rocks* from Greenland show that there existed liquid water and also living cells on Earth already 3.8 billion years ago. Although seemingly that in this early time period already temperatures for chemical reactions have been ideal, no chemical or even biological evolution could have put into effect at the surface of our Earth, *because in the lakes and seas must have been presented an evolution-hostile chemical equilibrium.*

However, as confirmed by analytical methods, chemical isotope signatures are controversial, since such an early origin of life is contrary to the chronological table of evolution. But perhaps, such primeval living cells could be a misinterpretation, because the structure looking like microfossils are in the opinion of other expert's *inorganic* compounds, which also occur today at hot deep-see fountains. Accordingly, another controversial finding is valued, which does not fit into the supposed very slowly ongoing evolutionary development after Charles Darwin. According to the geologic time scale almost 3.5 billion years old microfossils of the Apex Formation in Western Australia (Brasier et al, 2002) seem to appear much too early and occupy a much too high state of evolution to represent the true beginning of life – if the theory of evolution is taken as an indication. Although these microfossils (Oscillatoriacea) exist at the base of the evolutionary family tree of the cyanobacteria (blue-green algae), there must have been taken place an evolutionary development over long time periods in order to produce such a complex life with photosynthesis and this already before at the given geologic time of 3.8 billion years ago. However, the deposits appear to originate not from the bottom of a shallow sea, but from hot springs deep below the water surface. Thus, the assumption would be refuted, that these apparently microfossils are cyanobacteria, which use photosynthesis, because the light does not penetrate as deeply into the sea. These

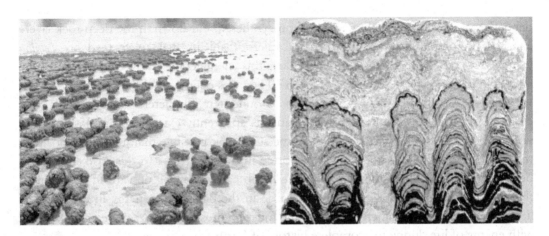

Figure 75: From 3.5 billion years ago? Left picture: *Living stromatolite in the marine reserve on the Western Australia's Shark Bay. Right: section of a fossil stromatolite from the time period before the Cambrian explosion (Proterozoic), when at this time there was no life on land. However, are such structures – as here to see from the Eastern Andes of Bolivia – the result of animal life or alternatively formed inorganic, resulting from self-organization structures? See Figure 53 on page 128..*

structures should therefore be of inorganic instead of organic origin. But nowadays, stromatolite – a laminated usually mounded sedimentary fossil formed from layers of cyanobacteria, calcium carbonate, and trapped sediment – are known as the oldest fossils: It seems to have existed already 3.5 billion years ago. Even these seemingly biologically formed stratified rocks are controversial. The same simple layered structures result from processes of self-organization which are *purely* mineral origin, will say that it is possible to develop such a fossil stromatolite through and through solely by systematically inorganic processes. Such lifeless structures looks like those of biological origin.

If you consider that earth scientists still have trouble agreeing on reliable bio-signatures, then it is understandable that reliable traces of former primitive organisms – such as on Mars or other celestial bodies – can hardly be identified unambiguously respectively without doubt.

Subterranean Life

On Earth, there are life forms on the surface, which transform light into energy using photosynthesis. But according to the foregoing explanation, however, life cannot be originate at the surface! But there is another area where life exists which do not dependent on the Sun or light energy (photosynthesis). Far down in the deep rock layers and the bottom of the oceans, there are microorganisms that solely feed on *chemical* energy. This raises the question of where does this energy come from? May be out of the atmosphere respectively near-surface layers? Then this energy would have to go through the layers of soil to a depth of perhaps around ten kilometers. This should be a continuous process, since these microorganisms apparently existed before 3500 million years ago corresponding to the geologic time scale. Is it possible that just the opposite is true and the energy need not to be transported into deep layers? In this case energy will be constantly available in the depths of the Earth and is flowing upward in direction to the surface through geological layers? Because in the depths life energy in the form of light is not available, another energy source must be present to provide microorganisms with energy of life during the complete history of Earth.

Not only by evolutionary biologists is usually represented that biological evolution is a proven fact. Therefore, all controversial artifacts which are discovered in to old geological layers have to be forgeries. As an example of the assimilation of organisms which are living on the surface using light energy are scientifically described those microorganisms which exist down on the ocean floors or in deep crevices and live on toxic chemical elements such as sulfur or hydrocarbons exclusively without oxygen (photosynthesis). Therefore this perception is the base of the officially statement that life which is originated at the surface can adapt actually hostile circumstances: a alleged proof of evolution. But this must be a wrong view, because studies show that the expressions of life in both regions – at the surface and in the depth of Erath – seems to trace back to a common origin, since both life forms are based on the same genetic system.

Surely, there have been contacts between the life on the surface and in the depths of the Earth, but from this one cannot derive any dependence of the two areas of life from each other. If we find a dependence of one region from the other, then life should be created in the independent sector and have then spread later on to the dependent. Since life on the surface depends *only* on the capacity for photosynthesis and the solely chemical energy processing microorganisms are a derivative form of life, the photosynthesis must have already developed at a very early stage of Earth history, respectively at a time period as creatures were able to assimilate only chemical energy which is existing all the time! Therefore photosynthesis must have been available on the Earth's surface without interruption since the starting of life. If this should be true, then this life would have to go with evolutionary steps into the depths of the Earth, almost in the form of an invasion. It would not only must be developed a skill to use the chemical energy source available there, because photosynthesis does not work down there, but microorganisms would have had also to brave the prevailing high pressures and temperatures down there at the same time. Is this scenario realistic? Hardly, because life on Earth may have had no repair mechanisms to correct the inevitable damage caused by solar and cosmic radiation in the beginning. In contrast, the living conditions under the surface are much better almost without troubles.

Already Professor Thomas Gold pointed out the advantages of the biosphere in the depth and assumed that the start of life most likely occurred in the depths of the Earth, and called this "the deep hot biosphere" (Gold, 1999).

Down there is no need for complex molecules to protect against radiation which are bombarding the Earth's surface. Although there is radioactivity in the depths, it penetrates to a much lesser through the rocks, as it is in point of fact at the surface of the Earth. In addition, the intensity of the radiation deep in the Earth's is very consistent over long periods of time. A fact that was sure not given constantly at the surface of the Earth's. Because there, life was over and over again threatened by natural disasters.

A large meteorite impact was even able to destroy life on the surface almost completely, as it should be happened not only at the end of the dinosaur era. In contrast, such a giant impact could have been beneficial to life in the depths of the Earth, because it would have formed new cracks in the stony layers. As a result more energy in form of hydrocarbons would be increased, and in this case the existing microbic life could grow inflationary. Life in the undisturbed depths of the Earth could survive very long periods of geological time, if there is sufficient supply of chemical energy.

In the depths of the Earth microbes bear high temperatures whereas these life would boil at the surface of the Earth. The reason is that the high pressure in the depths cause that the boiling point of water is considerably higher and therefore liquid water is in a much wider temperature range available than at the surface. This is favorable for microbes due to the low complexity, which in contrast to more complex organisms are able to survive going along with a larger temperature variation (see page 167).

In the depths of the Earth microbes also need not to fear bad weather, super-floods or evaporating oceans. Down there reign an uniform temperatures, while at the surface sudden temperature changes happen from time to time. Because water remains liquid at the relatively low pressure of the atmosphere, the temperature may not vary greatly through long geological periods of time, to sustain and maintain life. But this pre-condition exists in the depths of the Earth, apart from the environment of active volcanoes and other hot areas (hot spots).

We can say that life in the depths could have been developed relatively undisturbed over long periods of time and was hardly disturbed by external influences. The existing life in the depths of the Earth is universal without any photosynthesis-processes since the beginning of geological time. The requirement is that the chemical-to-use energy will be constantly available down there. If not, stable life cannot exist in the depths of the Earth. More precisely, this energy must rise from deeper zones, respectively from areas where life cannot exist, because there the conditions of pressure and temperature are outside the life range of microbes: Otherwise, they would have "drained" the sources of energy.

Because the energy sources are lying deeper than the living area of microorganisms, this anaerobic life was constantly supplied with chemical energies, particularly hydrocarbons.

Let us now briefly look to Mars, then it seems no longer incomprehensible that the Viking landers in 1976 there found no evidence of aerobic life. On the surface of Mars, it is hardly possible that there exist life, according to preceding statements. In my opinion, higher and more complex life is there not to find on the surface. The experiments performed for the detection of metabolic productions and in particular the detection of photosynthesis had to provide negative results. For life on Mars there is no necessary to develop a photosynthesis, because chemical energy is amply available.

As expected the Viking experiments have given no indication accordingly to biological activity (metabolism) and carbon-based organic reactions, but instead of this to inorganic behavior. In this way NASA interpreted the results of measurements, but it is doubted by many others up to date. Why other probes landed on Mars were not equipped with appropriate laboratories to detect metabolism (life forms)? Why scientists are searching exclusively for the existence of water nowadays – which I had already predicted in my first book published 1998? Because evolutionary biologists are convinced that water is the most important prerequisite for the development of life. But this opinion can only be correct in case of *aerobic* life which needs photosynthesis to generate energy. In contrast, in this way it is not possible to detect anaerobic life, which thrives without any oxygen and water because this life-forms needs solely chemical energy without photosynthesis and light energy or simply water. Although there is water at least temporary on Mars, but to supply evidence for photosynthesis are therefore likely not to become.

I could not figure out whether the experiments of the Viking landers ever had the opportunity to detect methane. At that time, in 1976, it had been declared insane, if you would have pointed to the existence of methane on Mars. Why do you get these days such little information about approved methane findings in recent years? Only to prevent speculation about life on Mars? No, because in the context of the reigning cosmology cannot be explained the mass of occurring methane on other planets and even comets and this without any biological processes.

But methane is the most common hydrocarbon and a main energy source for chemical life. This anaerobic life could be viewed as a precursor of aerobic life through photosynthesis. After the Viking missions, other space probes have provided evidence for chemical reactions and this show that Mars is not in chemical equilibrium. Therefore, on Mars should take place processes of self-organization, resulting from inorganic structures that resemble those of Earth. Because hydrocarbons are present on Mars, similar structures can there arise even in organic form. Self-organization takes place in the organic and inorganic chemistry everywhere in the same way. In other words, in the depths of Mars may life forms (microbes) exist, which live on purely chemical energy. But nobody looks after it, because, according to evolutionary biologists, life originates only in the water and live on photosynthesis, thus not solely on chemically energy.

But the case is reversed: Why eventually existing microbes in the depth of Mars should change their generation of energy towards photosynthesis, respectively light energy? The average temperature on Mars is –55 degrees Celsius. This is therefore outside the temperature range of the maximum transmutation probability, because on the surface of Mars is ruling a temperature maximum of +27 degrees Celsius, which is just in the limit range temperature for microbes. To achieve the ideal temperature at which it is possible to develop complex life-forms, it will be missing about 10 degrees Celsius. But in the interior of Mars, there is this ideal temperature, and it should be simple life-forms down there, but micro-biological life solely living on chemical energy.

Coming back to Earth. From the previous considerations we can develop the working hypothesis that life was not originated at the Earth's surface, but instead of this in the depth as the most likely place. The most important difference between the deep hot biosphere and the surface biosphere as realms for the origin of life is the abundance of primordial energy upwelling from below and just as important, from a depths so distant that the source itself is inaccessible for microbes. According to the deep-Earth gas theory, hydrocarbon fluids diffuse and migrate upward. But down there it's much too hot, as that microorganisms can reach the source of the hydrocarbons down there. Therefore, this source cannot "dry out" because this energy fountain cannot serve as a direct food source itself. But microbes existing in higher layers or in cracks and pores in rocks are supplied constantly with chemical energy over a long period. This energy, in the form of hydrocarbons, are migrating upward from energy sources in deeper layers.

"For this reasons alone, the subsurface is the more likely location for the early phases of life if the deep-Earth gas theory is valid. Hydrocarbon fluids streaming up from below, and originating at depths far too hot for carbon-based life to reach and plunder, might have offered sustenance at a steady, metered rate for long periods. This scenario would provide ideal conditions for life to arise and flourish. Later, the same conditions would have allowed microbial life to develop a range of chemical abilities. Motile adventures at the outer edge of the subsurface realm – perhaps in the vicinity of vents on the deep ocean floor – might have developed heat-sensing pigments by which to orient themselves toward the energy-rich vent, thus preventing their being carried away into the barren, cold ocean. It has been suggested that the first important step toward photosynthesis was orientation and navigation by the sensing of heat radiation. Photosynthesis would have followed if some microbes found it advantageous to live at or near the surface and to enrich their energy by the use of sunlight" (Gold, 1999, p. 168 f.).

The aforementioned principle of direction finding of microbes due to thermal radiation in the vicinity of deep-sea volcanoes could represent an early preform of the photosynthesis, as Euan G. Nisbet suspected (see Nisbet / Fowler, 1996). In principle, there is again an example of self-organization, because bacteria follow the higher concentration of attractant and in this case the microbial level of thermal radiation.

Hyperthermophiles, respectively excessive heat–loving archaea and bacteria are the deeply rooted and the most ancient, respectively the oldest life forms. It is thinkable "(though very unlikely in my views) that the surface of the Earth some three billions years ago was fit only for the most highly heat-loving forms of life. The deep–rootedness of extreme thermophiles might also be taken to suggest the hot ocean vents as the locus of life's origin. But what would have been the source of chemical energy, if no fluids upwelling from the deep that could derive energy in reactions with materials available on their pathway? Empirical evidence, then, though it does not exclude hypothesis that posit a surface origin of life, nevertheless strongly supports the concentration of subsurface origins" (Gold, 1999, p. 169 f.).

Going in the other direction, from surface to deep origin of life, Gold "cannot see a similarly favorable situation. Photosynthesis would have to develop, as a very early step, from an evolution dependent on some unspecified form of chemical energy becoming available continuously on the surface. The progression would then have to take the form of invasion of the subsurface. The development of an ability to use the chemical energy sources available there, and an accommodation to the elevated temperatures and extreme pressures of that realm. The subsurface realm, for a number of theoretical reasons, therefore appears to be the more likely site for the initial development of this curious and extremely elaborate chemical processing that we call life. Pressure and temperature conditions in the subsurface are steadier and generally more conductive to life than are surface conditions. Subsurface life would not only have tolerated large asteroid impacts but would probably have benefited from the disruption. Radiation harmful to life would have been greatly at depths. Finally, according to deep-gas theory, chemical energy would have been abundant and supplied as a metered flow" (ibid., p. 169).

The conventional views of the evolutionary biology, that life has formed in a location on the Earth's surface with a probability of about zero purely by chance, we had rejected solely on the basis of the chemical equilibrium in lakes and seas. That life then even had populated the inhospitable depths of the Earth seems almost absurd, because how many useless trials life was forced to start on-again-off-again? Was there ever life enough units to launch a suicide mission "invasion into the deep"? In addition, no intermediate forms have been recorded (missing links) in the development of the species (macro-evolution), as discussed in detail already in my previous books. The findings of fossils suggest that already at the beginning of life all animal phyla roamed on land, and exist to this day. Therefore, I had propagated the constancy of species and decided against a genealogical development – see in detail my book *The Human History Mistake* (Zillmer, 2005). "Darwinian selection pressure is too low to" achieve a new development of multicellular organisms", writes my friend Professor Dr. Wolfgang Kundt (2005, p. 207), University of Bonn in Germany. This problem is also known as "Darwin Dilemma".

There is no family tree, although all animals and humans are genetically similar. For example, a mouse and a horse differ totally purely superficial, but all vertebrates have the same types of cells and molecules! In contrast to the daily occurring micro-evolution is a process of macro-evolution with a slowly development of new large animals or even new organs not yet proven to this day. The reason is obvious, because there is no evolution in the sense of macro-evolution of various cells in the body of existing animals to develop new species; but what real happen are only changes to the arrangement of existing cells as well as the speed and number of the alternation of generations of cells, which are kept viable. In simple terms, nothing new is slowly developing, but existent genetic makeup is only newly combined, resulting in *directly functional* life-forms. *Darwin was wrong!*

The determination of the *Cambrian explosion* could have meant the end of the evolutionary hypothesis. For Charles Darwin himself wrote: "Could be demonstrated due to any complex organ, that its completion would not have been developed by numerous, successive modifications, so my theory would absolutely break down" (Darwin, 1859, p. 206). And it has collapsed, because many fossils exist with no changes or only little variations over long geological time periods before then a sudden change over happens to originate new shapes and sizes with no intermediate stages.

In the scientific journal "Science" (vol. 293, 20 Jul 2001, p. 438 f.) is confirmed, that the beginning of the Cambrian period (around 530 million years ago) saw the sudden appearance of nearly all major groups of animals (phyla) in the fossil record, which still mainly represent the current biota (Fortey, 2001). The British paleontologist, Derek V. Ager, admits: If we examine the fossil evidence in detail, whether the order (taxonomy) or the species level, emerges one point: What we find over and over again is not a gradual evolution, but a sudden explosion of a group (*Proceedings of the British Geological Association*, vol. 87, 1976, p. 133).

Intermediate forms were not found to this day, however, but a sudden appearance of new and more perfect animals without any intermediate steps. Such a scenario can be proved in the layers of the Cambrian.. This wonderful event 542 million years ago (on the geological time scale), started as a sort of evolutionary big bang of highly-developed animals and is known in the geological literature as the *Cambrian explosion* because in earlier stratums organic life can only found in the form of single-and multi-celled organisms which exhibit no skeleton and no hard parts. But suddenly it is teeming with complex life forms of all types, those have arisen without a long-term evolution and without intermediate steps, proved by perfect fossils without intermediate forms or any aberrations which are preserved in geological formations.

If we bear in mind that all animals are genetically close and seems not necessary to develop new cells for the appearance of new huge animal, but instead of this happen a change of the arrangement of the cells and the speed as well as the quantity of alternation of generations; then it is plausible that seemingly new species appeared which are intact

Figure 76: Pig monkey from China. Sudden species transformation without Darwinian macro-evolution.

animals already at the time of "creation". Thus, in 2008 Fengzhang (China) was apparently born a pink monkey. But his four siblings are healthy piglets. In fact, this ape-like creatures had pig's paws as the pig's siblings. This monkey-pig died a few days after birth from a cold. Sudden changes of new species without intermediate steps seems to be no magic.

Such may also be explained why additional functioning body parts can be formed directly. 1697 became an unknown Turkish archer in captivity, who had two heads. Aldrovandus Ulisse (1642) reported a monster that was raised at the Scottish king's court in 1490. It is said to have sung to the admiration of the audience in two parts. There are others besides the phenomenon of multi-breasted. A woodcut from the 19th Century shows a young mother with an additional mammary gland on the left thigh. On this was shown a toddler suckled up to 23 months, while at the same time an infant was breastfeeding at the regular mother's breast. Another phenomenon is the sudden gigantism. The greatest man to have lived should be Robert Wadlow, who died in 1940 with a size of 2.72 meters, followed by Grady Patterson with 2.65 meters and John F. Carroll with 2.64 meters.

Why it is possible that creatures emerge at a single blow which look quite different, but are fully functioning? Everything is apparently based on a single cell type developed by nature! Although there are various types of cells, but these arise from so-called stem cells, commonly referred to as somatic cells. These have influence depending on the potential to develop into any tissue (embryonic stem cells) or in certain tissues (adult stem cells). Stem cells can also generate daughter cells that in turn possess the properties of stem cells. About the fate of each cell decides it especially the biological milieu. The mechanisms are coming into play are not fully understood. These stem cells, which are referred to as a pluripotent, can differentiate into any cell type of an organism, because they are not set to a specific tissue type. However, even these are not in a position to form an entire organism. Life is apparently universal and is based at the same time on

Figure 77: Double-headed Turkish archer.

Sudden emergence of a new, fully functional species. A reproduction of this would require two identical individuals. This works usually not in case of mammals due to the small number of individuals and alternation of generation, but in contrast to this it is possible at the microbes-level, especially if it is a self-catalyst. Picture from: Holland, 1921.

certain, seemingly fixed principles. But these bring self-organization to mind and this is certainly not purely coincidental. In principle, only once at any given time a single stem cell must have been developed, which reproduced en masse. It then began a structural and functional specialization in the form of an development of individual cells, accompanied by a decrease in the cell division rate and the losing of to be an all-rounder-cell.

In contrast, may have been random mutations in highly developed organisms no reasonable mechanism for evolutionary change, because there is the body's own repair mechanisms, which eliminate random errors. After Charles Darwin errors in the genetic material are necessary, then called favorable mutations. But in this case it is required an exorbitant number of attempts. However, the number of living individuals of a given species per period of each alternation of generation is too low. Cannot be there other reasons for genetic variation as random errors that will affect favorable mutations? Is it possible that there be a process of self-organization for the emergence of life?

Organisms that occur in an infinite number, are composed of only one cell. For example, in the intestines of humans and animals resides Escherichia coli, short form *E. coli*, the most-studied bacterium on earth. These bacteria are known to be essential for the chemical digestion of our food. In the intestinal tract of every human there reside about one million times one million, so about 10^{12} such bacteria. Multiplied by the number of living people on earth, this results in an unimaginable number of perhaps 10^{22} E. coli bacteria. At normal conditions, a bacterium suitable fed will reproduce about every 20 minutes. With this unimaginably large number of options increase, it is conceivable that a microbial clade "developed" (synthesized) an enzyme or a complex molecule, which had a useful function in a metabolism.

We now compare this almost seemingly endless development possibilities with those of mammals. If abundant reproductive microbial strains may provide *one successive chance step in 100 years*, and this rate is transferred to macro-biological life forms, for example 100.000 elephants alive today with an alternation of generation of about ten years, then this would translate into one billion times one billion (= 10^{18}) years for the same chemical evolution in elephants, and similar times for any macro-biological form. In other words, because the bacteria reproduce 262.800 times as fast, and there are 10^{17} times as many in just the one habit (the human intestine) that we have mentioned then the chance of hitting on a favorable mutation in a given period of time is thus $2,6 \times 10^{22}$ times greater for these bacteria than for the elephants (Gold, 1999, p. 177 f.).

In addition to many empirical evidences and finds, the probability calculation is against any development of a favorable mutation in macro-biological life forms, because the need for only one small evolutionary step of one billion times one billion years will be out of any chance. In order to reach changes in the internal chemical processes there have to occur *numerous* of these intermediate steps; even there is *every* step to estimate as impossible. Therefore, random mutations of macro-biological life forms often lead to damage and not to further development.

Thus chance mutations and their selection could well be the path to major evolutionary innovations among the various lineages of microorganisms and yet be hopeless slow for large creatures. As it turns out, large creatures differ greatly in form, but no so much in function. A mouse may be look very different from an elephant, but all vertebrates share the same cell types, the same species of molecules. Metabolic innovation may have been the more difficult to archive. A methane-eating, heat loving microbe may look very much like a photosynthetic microbe, but the evolutionary differences are far greater than those that separate jellyfish and human. (ibid., p. 178).

In other words, a macro-biological development would never happen. We must now ask whether such genetic processes occurring in nature. According to Darwinian logic such processes must appear to be highly unlikely. But it can happen if it is possible to "bring a molecule from the microbes into the genetic material of macro-organisms in an acceptable time frame for evolution" (ibid, p. 179).

In this way microbes take over the chemical digestion in the digestive tract of every human. To act in this way, the microbes possess this ability already before there existed humans. Therefore it has taken place *no* macro-evolutionary development in humans itself, as Charles Darwin imagined. But nevertheless it is possible that the already well-functioning bacterial clade continue to develop furthermore *within* another organism, as already described.

Major innovations in terms of metabolism have been almost fully achieved in the area of life from microbes, such as convincing empirical evidence and theoretical arguments by Lynn Margulis show (*Journal of Theoretical Biology*, vol. 14, No. 3, pp. 255–274). She developed the already by the German botanist Andreas Schimper (1883) postulated and by Konstantin Merezhkovsky (1905) again proposed endosymbiotic theory, which assumes that the today's complex cells were composed of less complex components, and finally the chemical evolution (!) build the starting point of the origin of life.

Furthermore, this theory states that the cell of a unicellular organism was incorporated of another protozoa and so became a part of the cell of a higher organism which was suddenly arisen. In this way more and more complex organisms were generated. Also components of human cells trace back to unicellular organisms. This theory was initially dismissed and is ignored up to date.

Let's go back to the beginning of the development of life. The endosymbiotic theory states that at an early stage of development within the range of eukaryotes (in which all living higher organisms with a nucleus, cell membrane and several chromosomes are grouped together), so-called cellular organisms (bacteria and archaea), which do not have a true nucleus, have absorbed a biological membrane (endocytosis) into a precursor proto-cell. In such a way they developed into organ cells. In other words, the protozoa entered into a so-called endosymbiosis with an eukaryotic cell and lived within their host cell for their mutual benefit.

Incorrectly, this endosymbiotic development of life is going to be understood as an example for evolution according to the Darwinian theory, but it should be clear that no imperceptibly slow development of macro-biological life forms happened corresponding to the evolutionary theory, because in fact there is a quick step in the development of the host organism. Thus, the principle of suddenness – postulated in my previous books – is confirmed in contrast to a barely noticeable Darwinian evolution! It is also confirmed, that a better result can be achieved due to cooperation instead of Darwinian confrontation (Zillmer, 2005, p. 302 f.). Darwin was wrong!

Now one could imagine that more and more complex organisms has been built up caused by rapid leaps almost without any intermediate steps, on the basis of a certain principle which was once successful. And if there were intermediate steps, these were *already functioning* not later than at the time when they became a part of the higher developed organism. In contrast, Darwin's idea can not to be true if we act on the assumption that there happens always an excruciatingly slow development, because only partly developed and therefore not yet functional organs or limbs convert an animal into a stupor, which will be eliminated after Darwin's theory.

Added to this, perhaps other mechanism happened which allow a development of fully designed principles with rapid steps *without* intermediate stages, namely the gene transfer from one organism to another (see Miller, 1998). These are available for viruses and is known as horizontal gene transfer. It is known that bacteriophages (short: phages), a specific group of viruses, inject their genetic material into host cells, whereby bacteria and archaea thus be greatly changed. For example, nitrogen-fixing *azotobacter vinelandii* shows a completely different shape and kind of flagellums and cell cyst after taking a complete transformation of *bacteriophage A21*. Such cells, which differ so greatly from the original organism were primarily believe to be a totally other organisms. However, it was found out by now that a new species seemed to be present only as long as there the foreign genetic material (phage genome) was inside of the host cell. The process of change could be reversed afterwards and was therefore reversible. But in the case of such a transfer of genes no new genetic information's will be generated, but already existing genes are arranged new between different individuals. Therefore micro-evolution happens but no macro-evolution, because nothing universal new is added.

If in the nature genetic material can be transferred from one genus to another, then the coexistence of different creatures gets a crucial importance for the mutual advantage: cooperation instead of confrontation.

Today one can observe different species between symbiosis and endosymbiosis in various stages, such as the roots of some plants live in symbiosis with nitrogen-fixing bacteria or plant lice in symbiosis with algae or bacteria. Could such colonies have also communicated their genetic material? For this case, already a new complex multicellular organism originate itself, without any macro-evolutionary interstage. Already Vilmos Csanyi (1989) has proposed a symbiotic fusion of different microbial clades as the best explanation for the fundamental differences between the cell types in animal bodies.

Mutations can be successfully performed only by microbes, if we take in our mind the unimaginably vast number of entities and the speedily rate of reproduction. But this is not true for higher developed (large-scale) life-forms, as it is confirmed in the German time life-book *Evolution* by Ruth Moore (1970, p. 91): "The work in many laboratories showed that most mutations are harmful and if those are drastic changes they are fatal. They suggest a certain extent in the wrong direction, in the sense that any change in a harmonious, well-adapted organism has an adverse effect. The most profound mutation carriers do not stay alive long enough to pass the changes to their offspring."

If one is going to perform unplanned change in a robot or a specialized organism, the mechanism is certainly *not* upgraded, but in all likelihood damaged, or, at the best, shows no effect. A complex mechanism, which shall live, must work immediately. Somewhat simplified, one could compare the principle proposed here with an industrial production line in modern car factories. There, already functioning components – which are complete constructed outside of the installation site – are integrated into a system in form of a quick-accumulative process to function directly; comparably to the endosymbiotic theory.

Leaps in evolution will not be carry out with the development of different cells of larger animals, but instead of this with the arrangement of these cells and in the speed as well as number of the alternation of generations of the different cells, which are maintained. Microbes protrude out of here by the diversity of their metabolic capabilities. Endosymbiosis and gene transfer are manifold ways to allow sudden leaps.

If the before described processes and scenarios proceed, "then we should not describe the vast diversity of life through time as an evolutionary tree, with each branch progressing on its own and developing into individual species. Rather, we would think of a combined evolution of terrestrial biology, all continuing to be closely interrelated with one another and with the most prolific gene pool of all – that of the microorganisms. Because the deep hot biosphere is, in my view, so vast, and because this realm is very likely to have natured the first living systems, many of the innovations and gene-trading and merging events that support today's expressions of life probably took place well before there was any life on the surface of the Earth" (Gold, 1999, p. 183).

Therefore one could argue that without microorganisms there are no complex life forms on Earth possible. Life evolved in the deep hot biosphere and the initial point of development was a starting chemical self-organization process. Curiously, we had already discussed inorganic structures that confusingly appear similar to the organic. These are *two different directions* of development, but both have been shaped by the same principle: self-organization. We can therefore not expect that from inorganic structures arise suddenly living respectively organic structures, because all known life forms are carbon-based. Therefore it is to differ organic and inorganic chemistry and inorganic structures cannot be alive. But organic life can solely live on inorganic energy – particularly this was the requirement to enable the start of life.

Accordingly, it is not possible to identify a single step that separates a stage of living matter from the structurally similar but non-living inorganic matter. Therefore, a connector cannot exist between non-living and living matter, because the development of life have occurred in the range of microorganisms. The living cell arose not suddenly and randomly, how it can to be show with the *statistical chemistry*. The biologically active substance consists of compounds with carbon. Therefore, the ancestor of life should be a carbon mineral. For this purpose, graphite fits as a stable form of carbon, because it can absorb many different materials and will be chemically active (Oesterle, 1990).

The intensive research of W. J. Sawenkow (1991) have shown that the biologically active cell is caused by self-organization of filamentous graphite crystals with screw-type dislocations, around which have formed a casing to be composed of organic materials.. The distance between the turns in the graphite crystal and such in the DNA filament is approximately the same. The famous Austrian scientist Erwin Schrödinger (1951) believed that the genetic information could be encrypted into something like an "aperiodic crystal", which is able to build up again and again.

Nature has selected only 20 amino acids to build basic modules of proteins, which provide enough combinations and deliver all proteins necessary for life. These chemical and physical processes require energy, which must be supplied. As previously described, Stanley L. Miller tried to reconstruct this with a mixture of simple chemical substances in a hypothetical primordial atmosphere. This was exposed by electric discharges, which should simulate lightning. After this procedure it was possible to account for amino acids and other substances – basic modules of biomolecules – which were originated.

Although, at the surface life cannot come into being. The electrical discharges are interesting and in the analogous simulate the conditions of a cold-electric Earth because in the depths happen discharges respectively flowing electrical currents. Because the substances, which Miller mixed together, are also present in the depths of the Earth, it is possible that there amino acids could be originated.

Also on comets could run these processes, as between the cold-electrical primal Earth and the comet is no essential difference, except that the pressure and temperature conditions are somewhat different. Therefore, it is not surprising that without an atmosphere, like subsurface of the Earth, in comets also originate amino acids, because there exist the same chemical substances used by Miller for his laboratory experiments.

The already 1864 touched down meteorite Orgueil was examined again in 2001. Pascale Ehrenfreund of the *Leiden Observatory*, Netherlands, studied a relatively simple mixture of amino acids that not ended up in the rocks due to a terrestrial contamination. These results were then compared with the present studies of three other meteorites: Murchison, Murray and Ivuna. The first two contained a very complex mixture of amino acids while in Ivuna and Orgueil mainly were detected two different amino acids.

Also in the universe suchlike have been discovered: glycine (amino acetic acid), the smallest and simplest protein-building amino acid (Belloche et al, 2008). Glycine is an important hub and an important component in the metabolism of nearly all proteins.

The role of electrical energy is often neglected, whether in Miller's experiments, or the formation of components of life, or the functioning of complex biological compounds – but also in geology. So it is possible with naturally related electrolysis experiments to generate rhythmic mineral structures, such as bandings, which so far could not acceptably explained to form through gravity or sequential substances supply (Jacob et al., 1992). In other words, in the laboratory can be produced geological bandings or strange mineral mixtures with the transmission of electric current in a *few days*, for which would be need millions of years to build such banded geological layers according to the conventional geologic view and time scale.

The pursuit of chemical open systems for maximum stability, even called conservative self-organization, is the driving force of development. So not only natural mineral deposits but also bandings are formed in the lithosphere. This happen analogical to periodic structures in the physical chemistry as a result of internal self-organization.

Thus one can argue that life on Earth is originated according to physical laws, for example like minerals. Nobody claims that minerals must be imported from space to Earth to come into being. Nor is life on Earth originated in outer space. Since the origin of life on the one hand can be described by self-organization and the laws of micro-evolution, and on the other hand a slowly performing macro-evolution of complex life systems in accordance with Darwinian evolutionary theory is not proven up to date. I would like to introduce a new term for the previously outlined origin of life through self-organization and call this principle "autogenously biogenic-organization" (shortly *autobionization*).

How quickly life could had been developed? The calculus of probability leads us within the range of an unimaginably numerical data, in which it was possible to build the *first* magic molecule within a many number of useless companion. This molecule alone cannot achieve anything, unless it is able to reproduce itself quickly, so in order to settle the entire "primordial soup" in a short time period. This is the case, if this molecule was an oneself-catalyzer. A normal catalyzer caused only the formation of any other molecules, but a oneself-catalyzer can be made copies of itself repeatedly from chemical matter of the "primordial soup".

Because the reproduced copy is equal to the original, this new molecule also works as an oneself-catalyzer and can also bring copies into being. Since these copies produce copies again and again, the number of molecules grows exponentially rapidly. After a day there are perhaps two, after two days four and only 30 days later on it should be possible to account about a billion. After 60 days there are theoretically more than 10^{18} molecules, that is one billion times one billion.

Of course this process is limited, occurring on the one hand by a lack of supplies of certain elements or atoms and on the other hand by the maximum pore volume in the rocks of the lithosphere. So that microbes are able to live subsurface in the rocks, it must be interspersed with cracks and fissures.

Antiquated Geology Textbooks

The deepest hole was drilled into the Earth's crust in the north of Russia, reaching a depth of more than eleven kilometers. Even at this depth running hot, mineral-rich water in cracks of granitic bedrock was found (Kozlovsky, 1982).

In Germany, they wanted to go still deeper, but at least ten kilometers, and started the deep drilling program of the Federal Republic of Germany (in short: KTB) near the town of Windischeschenbach. Scientists belief that in this area the primordial continental plates of Europe and Africa bumped together. Therefore, in this area shall have been exist a mountain range with the dimensions of the Himalayas 320 million years ago. Afterwards this mountain range was allegedly ablated. However, it is just purely a conceptual model, which is due to plate tectonics. There is no evidence for the existence of some kilometers high mountains in Central Europe. One can explain very easily the present-day geological formations by an expansion of the Earth and this without the belief of very high primeval mountains as a result of bumping continental plates due to plate tectonics.

However, on 12 October 1994 the German borehole had to terminate ahead of time because all geological expectations and predictions were thrown overboard. One goal of the project was to proof the supposed 320 million years old seam between the tectonic plates failed. Moreover, overabundant salt-water flowed in the borehole even at a depth of about eight kilometers, while at higher altitudes always water was streaming, to the astonishment of geophysicists: One normally thinks that the increasing weight and therefore pressure due to the height of overlying rocks are going to close the cracks down, so the hole should be dry in about nine kilometers depths, but it behaved exactly the opposite, stated Karl Fuchs in *Science* (Vol . 266, 28 Oct 1994, p. 545).

Because this water contained also metallic minerals, this discovery contradicts the view of many geophysicists due to the picture of the geological formations. Also the geological view, how far down cracks are possible to occur in the bedrock had to be changed. The drilling results are officially commented in this way, that there were received all of the desired results earlier than expected. But the opposite is the case, because the controversial results are observed only in silence and seldom are going to be discussed. Peter Kehrer was cited in 1994 (ibid., p. 545): The knowledge obtained by the deep drilling program indicates that the geology textbooks must be rewritten.

One goal of the project was to reach the zone where the rocks shall be transformed from a fixed structure into a fluxionary, in case of high temperatures and high pressures in the depths of the Earth.. The KTB researchers had not expected to be found a "flexible-brittle transition" in this depth, below which typically no earthquakes occur, before reaching a depth of ten kilometers and a temperature above 300 °C. Therefore, the opinions are divided as to whether already in 9.1 km depth at a temperature of "only" 280 °C, the beginning of the transition phase was reached.

Proves for a shifting of tectonic plates in the form of soft, water-bearing minerals (fluids) were not provided. Even a moving above the area where the fluid pressure shall generated cracks – which would suggest ductile rocks and move scenarios in the transition zone – could not be found. In fact, it was forced to terminate the borehole because the hole closed again and again itself (Kerr, 1994). A continuation of drilling deeper the borehole was not technically possible!

It is to ascertain that salt-water has flowed in depths of about eight kilometers and not until nine kilometers deep the granite transformed from a solid to a plastic state. There are different opinions among experts, but it could not has been confirmed the predictions of the alleged behavior of tectonic plates at that depth. The plate tectonics hypothesis has been put in question definitely!

If, as previously cited, the opinion is that the geological layers are squeezed together with the depth due to increasing pressure, so that eventually all the pores are closed and begin vividly because of the higher temperature, these layers are, then starts from very false premises. Actually agree to the match at about 9000 meters depth conditions

Figure 78: A schematic showing pressure regime in rock and fluid– filled pore-spaces. "*The height of a fluid-filled domain is limited since the rock will support only a limited pressure difference. For fluids less dense than rocks, it will fail in compression at the bottom and in tension at the top of a domain that exceeds a certain vertical dimension. (Rock is very week in tension, hence fluid pressures can hardly exceed rock pressures.) Domains of this critical height will migrate upwards*" (Gold, 1987, p. 83).

Ps = *critical pressure in rock.*

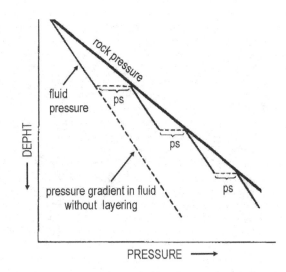

found here with the underlying theory, the gas sources requires even deeper.

If we consider, that water is still flowing in about eight kilometers deep, then the volume of rock pores corresponding to the density of water cannot continue over a certain depth beyond, because the rock has a limited strength. If the brake strength is exceeded, the rock will be yielding and the pores close, because the rock flows plastically. This is the exactly proven state in the depth of the German borehole in nine kilometers depth! At the same time the continuity of the pore pressure system, thus interrupted the interconnected pore space area, and there no fluid is flowing any more (Fig. 78). Up into this region the borehole was drilled at Windischeschenbach, and is was not possible to continue the drilling.

The height of such an (geological) stockwerk depends on the density of the rock, the density of the fluid in accordance with the ruling pressure, the strength of the rock under compression down there and of course the prevailing temperature. For methane and hard crystalline rock, this height is, under normal circumstances (temperature gradient) between four and eight kilometers.

If liquids would come only from the top down into the depths, then there would be no porosity anymore deeper than the critical depth, and therefore liquids have no network of fissures (cracks) to flow. If, however, fluids exist far below this critical layer again, then the situation is totally different. This scenario has already been described in detail and justified by Thomas Gold (1987, p. 81 ff.).

After the pores at the base of such a stockwerk are collapsed, like at the end of KTB-borehole in a depth of 9100 m, this area resides now *above* the supply source and forms a

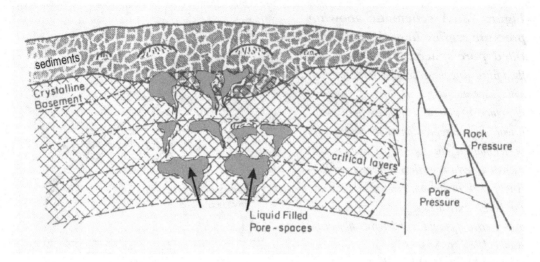

Figure 79: Fluid-filled domains in a vertical succession of pressure-regiments. The *uppermost critical layer may be in sediments or in basement rock. Low permeability layers in the sediments may provide islands of caprock (cover plate), retaining a reservoir below, and these were the formations chiefly sought in the search for oil and gas. The critical layers must be continuous surfaces — though in the possibly often far from flat (Gold, 1987, p. 85).*

kind of impervious cover plate (caprock) above the fluids down there. With a continuation of such upward migrating fluids (gases), the pressure increases with time in the caprock and their new fracture porosity will be originate or reactivate. With the time, the pressure, generated from the stockwerk below the caprock, will be larger than the pressure which results from the layers above. In other words, the upward-migrating fluids (gases) break open the previously tight respectively no pores containing caprock.

Subsequently fluids are able to stream upwardly through the critical layers into the geological layers above. But if the fluids are upwelling very rapidly through the caprock then the pressure in this layer itself falls off very rapidly at the same time, and the pores close again to create a newly tight layer (caprock). Then the described process starts anew until sufficient fluids will be accumulate again underneath the cover plate. Because the upwelling fluids generate additional pressure in the higher positioned stockwerk it will be create new porosity in the next higher pressurized caprock (= critical layer). Taking into account the strength or materials we can see a simple rule: In the moment, in which an area (stockwerk) spanned more height as it is possible with respect of balance and stability reasons, the fluid is sublevel-upwelling like a cascade towards the next caprock or at least the planet surface, *while the rock remains in the same area (stockwerk).*

Because the pressure gradient in the connected pore spaces is necessarily different to the pressure gradient in the surrounding rock, these liquids can form only a limited height interval.. The liquids are then in steplike arrangement with increasing depth. Each geological layer contains 'free' water, which the drilling engineer on the natural gas fields west of Ft. Worth, Texas, described to me as "formation water". The reason for this is to found in the strength of materials, because the liquids have to approach to the pressure gradient of the rock with a series of pressure sublevels (Fig. 79). But this can only happen if there are some separate areas with formation water.

The continuous rising of fluids (gas) in form of an upward-cascade is the only one non-explosive process of outgassing that is possible from deep levels and through solid rock. Only in this way can be fill-up successively one above the other area and this again and again, more or less slowly. So it can be explained the fact that exploited gas and oil fields periodically regenerate.

With the influence of the upcoming evolutionary theory the geologists of the 19th century accepted the principle of uniformity and developed a new geological view of the world. But at this time nobody knew that liquids occur deep below the Earth's surface in huge quantities. It was therefore assumed that only the strength of the rock shall prevail. Today we know more, and every civil engineer graduated in engineering geology, can surely reproduce the scenario described here, taking into account the strength of materials, if there subsurface exist liquids or gas wells.

Now the objection could be made that the rising methane would oxidize because the rocks contain oxygen, so that methane cannot rise vertically over long distances. There are also corresponding calculations, but these act on the assumption of a *small* amount of methane. Under this assumption, the objection would be justified, but it behaves differently, because a process of upwelling fluids which breakup cracks in non-porous rock requires a large and not just small amount of gas. The amount of oxygen, along the even free blasted crack-ways and rock surfaces, which adjoin to the gas-filled cracks is absolutely limited! If the oxygen – which is present in these areas – has been used up for oxidation of rising hydrocarbons there cannot occur further oxidation, and subsequent gas migrated unchanged respectively without oxidizing to higher levels and so on.

This process of upwelling fluids (gas) may be the explanation for different catastrophic events that occur in time intervals. If gas penetrates into a not active lava volcanic pipe so it is possible to enable a violent activity. In fact, during volcanic eruptions can also be detected methane. However, most of the gas is oxidized on the way up before reaching the Earth's surface, and therefore often occur the oxidization products hydrogen and carbon dioxide. Therefore, usually unnoticed and without comment, huge clouds of water vapor are escaping out of the volcanic vents. This could be observed in the first days of August 2001 at Mount Etna in Sicily. Officially scientists pass over this fact, because the conventional geophysics has no convincing explanation for the occurring of water vapor clouds .

Gas eruptions which are associated with earthquakes in the form of sudden precursor phenomena are often registered by animals and humans, but not by geoscientists who wait at this times far away from the event location to obey the swings of their seismic instruments only. Gas eruptions are also responsible for sudden activity phases of mud and lava volcanos. For this reason such activities are often accompanied by earthquakes and also associated with electric fields in Earth's crust, which are discharged from time and time. Therefore it is possible to see lightning's above volcanic eruptions or coming out of volcanic vents. Also, electrical sparks can ignite sputtering gas, so that sometimes the hydrocarbons go up in flames why it is possible that occur high fire tongues shooting up in the atmosphere, such more than 2000 meters high like in Baku as the already described before (see Picture 10).

With the temperature gradient within the Earth and the diffusion of electrons from inside the Earth (Thomson effect), an electric field occur at the surface (Oesterle / Jacob, 1994). The kimberlite pipes, with diamonds loaded vertical vents, may therefore be caused by short time overloads of this thermo-electricity, whereby can occur sudden, impulsive pressure changes. This ascent channels come into being where extreme heat accumulation and steep temperature gradients are mostly developed. "Strong magnetic anomalies confirm the electrical origin of the kimberlite pipes" (Oesterle, 1997, p. 99).

If there are gas eruptions under water, they are recognizable by rising bubbles. In dry areas, it is possible to obey that intensive emissions of gas set in motion loose sand. But natural gas does not always continuously streaming out like a breeze, but sometimes exploding suddenly or repeatedly. Such a sudden gas discharge caused earthquakes (not vice versa!) in Sumatra and the terrible 2004-tsunami in the Indian Ocean happened.

"Submarine gas eruptions are known from the Caspian Sea, off the coast of Burma and Borneo, on the coast of Peru and the Gulf of Paria between Trinidad and Venezuela. At the coast of Baku at Bibi Eibat submarine gas wells sometimes erupt with such sudden violence that boats already capsized when they came too close to the vortex. While calm weather, such gas eruptions are visible from afar. On the southeast corner of Trinidad has been observed submarine gas explosions, which threw up columns of water, accompanied by pitch and petroleum" (Stutzer, 1931, p. 280).

But sources of natural gas can also burn for years, like the eternal fire of antiquity. The already by Herodotus mentioned chimera (chimera) of Lycia allegedly burned two to three millennia. The burning gases of Nineveh and Babylon in ancient times were widely known. The fires of Baku, situated near methane emitting mud volcanoes, have led to the temple constructions of fire adorer. Even the fire cult of the Persians should be due to burning natural gas. The Zoastrians in Persia (modern Iran) pray at fire temples but they never worship fire because it is the symbol of their god Ahura Mazda (means *Lord* and *Wisdom*). The old but lost religion zoroastrianism was founded some time before the 6th century BC and has influenced Christianity, Judaism, and Islam. Because of drilling and the associated release of gases these fires are gone. Over the centuries, many exit points of burning gases have moved. This is an indication that gas is migrating upward from greater depths and search different paths with less resistance through porous rocks, cracks and sediment layers.

Life in the Depth

The formerly disputed, but in the meantime with the deep drilled boreholes in Russia and Germany proved point of view, that there water is still circulating several kilometers deep in the Earth shows that deep down are still exist pore spaces; this in areas where should be no pores and cracks anymore according to the official geological view. Based on undeniable facts, Thomas Gold (1999) formulated the theory of "deep hot biosphere" for which there is life up to depths of about ten kilometers.

According to the *deep-Earth gas theory* hydrocarbons are migrating upward in liquid form and penetrate into the pore spaces of Earth's crust, where is created a *primordial soup* and this shortly after the formation of our planet. Microbes can live up to a depth of about ten kilometers, because it is too hot for life still below this level. These tiny creatures can easily tolerate temperatures in the depth at which water is already boiling at the Earth's surface, because the boiling point in this depths is considerably higher due to the high pressure down there. Therefore, liquid water in the depths is available for life-forms in a wider temperature range, and bacteria which have low complexity can such use a range of approximately $\pm$ 20 degrees Celsius in contrast to humans.

At higher temperatures all chemical reactions go faster, and therefore in the soup down there occur a lot more compounds than at the surface. The volume of the habitat of microbes therefore includes the total pore volume in the rocks of the Earth's crust down to a depth of about ten kilometers.

We can now estimate the total pore spaces, which make up perhaps one percent of the estimated rock volume of 5.1 times 10^{18} cubic meters. The bulk of these cavities filled soup would be about 5.1 times 10^{16} tons. Let us assume that the average molecular weight of this soup would be at 50 units of atomic mass, then calculates Gold (1999, p. 171 ff.) about 6 times 10^{44} of such molecules. If each molecule once a day has the chance to go through a chemical transformation which can be triggered by heat, the coming together of the reacting molecules and / or additional (electrical) energy, then this is due to 3.65 times 10^6 variations of each of the 6 times 10^{44} molecules in a period of only 10.000 years. Overall, the mathematically result is about 2 times 10^{51} transformations. This is an unimaginable large number with 51 zeros.

This calculation is of course speculative, but in contrast to a development at the surface – due to the large volume of soup down there – this is a huge number of possibilities to form a certain single molecule. If this is a self-catalyst then its copies would be dominate the primordial soup relatively quick, as already described. This also increases the chance for a further development of this clade exponentially steep almost to infinity. But this process of life as a result of self-organization could only start and happen over long time periods, if a constant supply of energy and the flow of fluids from deeper layers take place, to those wells life cannot get down. In other words, the *deep-Earth gas theory* is a necessary condition for early formed subterranean life-forms.

The range down to about ten kilometers depth must be take account of the origin of life, alone to reason by chance. Is there evidence for the existence of microbes in this depths respectively for life without sunlight below the basement floors of our homes? In spite of hostile conditions, the depth is not lifeless. In recent years geologists and microbiologists isolated countless microbes from mines and drill cores that thrive miles deep subsurface and live on hydrogen, minerals, hydrocarbons (mainly methane) or carbon dioxide.

In the *East Driefontein Gold Mine* (South African) geologist Tullis Onstott of Princeton cataloged numerous species at a depth of 3.5 km and at a temperature of around 65 °C. The depth record currently hold bacteria that came from a gas well in the Swedish mountain from 5278 meters depth and there lived with about 70 °C.

As we visited gas fields to the west of Ft. Worth, Texas, I discussed with local experts problems of gas and oil drilling and was surprised by a problem not known to me to hear, but it was a decisive aspect to write this book. Although there in Texas is drilled mainly for gas to a depth of over 3000 meters, one would always hit petroleum. Thereby the drilling company is faced with the problem that sometimes clog the borehole completely at the deep end opening. Debt are microorganisms that apparently live on crude oil and quasi "swim" in it. Are there too many of microorganisms in the oil-containing layer, then the oil is too thick and no longer can be extracted.

Only the temperature seems to border the spread of microorganisms downward in the depth. The toughest protozoan are able to grow up to temperatures of 113 °C, as the in Regensburg, Germany, living microbiologist Karl Otto Stetter stated. This corresponds to approximately eight to twelve kilometers depth – depending on the density of the Earth's crust. "To this depth we suspect life in the deep hot biosphere", so Martin Fisk, geologist of the Oregon State University (*Focus*: 34/2000).

Deep boreholes are expensive, and to this day none have been made to be out for the mysteries of life in the Earth's crust. Thus, the exploration of the deep hot biosphere is only just at the beginning, as the mysterious microbes have been ignored long time. They produce methane among others. The Swedish microbiologist Professor Karsten Pedersen, Goteborg University, says that microorganisms stood billions of years ago at the beginning of life in the depths. This ecosystem is so new and diverse as it was the tropical rain forest 200 years ago for the Europeans, the biologist is convinced. He conducts research in a microbiology laboratory in Sweden, 460 meters below sea level, near Aspo, some 300 kilometers south of Stockholm. So far, the Swedish researchers discovered more than 200 different microorganisms. Some can breathe iron which is bounded in granite and thereby these microorganisms dissolve the rock. Others live on sulfur and produce toxic hydrogen sulfide gas. So–called methanogens build methane from hydrogen and carbon dioxide while an acetogen produces acetic acid from the same feed material. All microorganisms sucked out of the Erath and disperse over the entire surface there can be form a slime layer 1.50 meter high.

Pedersen is convinced that life could not get from the surface deep down into the Earth – for example with flowing water –, but as we have already formulated vice versa he ask: How did life come from the depths of the Earth upward to the surface? Living on methane, heat-loving microbes resemble outwardly such microbes which live on photosynthesis. Perhaps the area around deep-sea volcanoes is the place where life has started because there in the deep methane is upward-streaming.

Up to date, however, it is disregarded that electric fields exert a significant influence on the evolution of life. After the First World War, the Prague electricity corporation electrified crops during spring and summer. The yield was much larger than in normal fields. The surplus amounted to 50 percent for wheat and oats to nearly 100 percent.

At the end of the eighties at the pharmaceutical company Ciba-Geigy AG (now Novartis AG), the Swiss scientist Dr. Guido Ebner and Heinz Schürch made in laboratory experiments a sensational discovery, which was patented. They exposed fish eggs and grain to an electrostatic field in which no current flows (Fig. 45, p. 109). For example it developed corn into a prototype of itself, as such once sprouted in South America, with up to twelve instead of three pistons per stem. Without the use of fertilizers or pesticides were the growth and yield of maize increased significantly. In case of wheat there was growing a whole wheat bush with several ears of corn from a single grain of wheat, and this so quickly that the grain was ready for harvest after only six weeks. The company Ciba-Geigy AG suppressed the research immediately, because the sale of pesticides would have collapsed.

Interesting for our consideration was the trial, eggs of normal rainbow trouts to suspend to an electrostatic field. It came into being an almost extinct giant trout species that was a third larger, equipped with red gills and exhibited a distinct salmon hook on the lower jaw.

As the news agency *Associated Press* reported on 24 May 1989, were electromagnetic waves, probably gamma rays, responsible for a huge growth in plants. This occurred in the vicinity of nuclear power plant in Chernobyl in Ukraine, a while after a meltdown and explosion in the reactor core.

Electrostatic and electromagnetic fields can have a "special creative" effect on the development of plants and animals or micro-evolution, so that new life forms in case of "species transformation" can occur *suddenly*, sometimes even in opposition to the (apparent) "trend" toward to "archetypes" – without Darwinian macro-evolution. What does it means for the development of life, when the intensity of the "dipole field" of the Earth has dropped by half in the last 2000 years, supposed due to measured values of the strength of the Earth's field (Berckhemer, 1997, p. 146)? Is there a relationship of the formation of species with the strength of electric and electromagnetic fields?

Was the start of life in general only set in motion by electric fields? If microbes "invent" life in the depths and they are existent down there since that time, then there must be a permanent "gushing" source of chemical energy that is positioned below the life range of this tiny creatures. Only then the supply of energy could be ensured over long time periods, while at the same time remained a persistent chemical imbalance, so that a constant process of self-organization can occur. On the one hand this circumstances provided the precondition for the start of life ande care for the continuous renewal of "fossil" energies on the other hand. Is there evidence of a deep source of hydrocarbons?

If microbes "discovered" the life *in the depths* of the Earth and it always existed microbes down there, then there must be a permanent source of chemical energy whose fount line must lay deeper than the area of life of this tiny microorganisms. Only then the supply of energy could be preserved over long time periods while remained a continuous chemical imbalance at the same time. Thus, a process of self-organization could start, which provides the prerequisite for the start of life. Otherwise, if there was a chemical equilibrium, would be microbes and in the end also humans probably should not be originated! The question of whether hydrocarbons migrate upward from greater depths and creating a chemical imbalance is interesting not only in terms of our energy supply, but also for the beginning of life. Is there evidence of this deep source of hydrocarbons?

6   The Gas Wells in the Deep

Actually, cannot be stressed strongly enough that oil does not provide the composition picture expected from modified biogenic products, and any conclusions drawn from the constituents of old oils fit equally well or even better to the idea of an primarily hydrocarbon mixture, which organic leftovers are added, stated the famous chemist Sir Robert Robinson, president of the Royal Society in London as early as 1963 (Nature, vol. 199, 1963, pp. 113).

Old Beliefs

Hydrocarbons can occur at temperatures and air pressures which prevail near the surface in solid (coal), liquid (oil) and gaseous form (natural gas, with methane as the largest component part). Oil contains a wide range of different hydrocarbons, but have shared features. Therefore, one can suggest a similar mode of origin.

The scientists of the Western world claim that the origin of oil is clarified: It is caused by deposits of biological residues that have affected the geological processes. Therefore would have been countless alchemists feats in past geological eras, because oil is found almost everywhere on Earth. Thereby shall dominate high temperatures, but not too high, and increased pressure, as it prevails in the depths of the Earth, with a sufficiently long duration to ensure that the biological remains buried by soil deposits, possible can be converted very gradually. Thus for example, petroleum was described as fossil fuel and therefore to be limited in stock. This seems to be a matter of certainty, although the proven reserves from year to year not decline. On the contrary, the proven petroleum reserves are rising permanently, despite very low search activity. I remember the three car-free Sundays in Germany in the year 1973, when it was predicted by the Club of Rome the end of all petroleum reserves no later than the year 2000.

In contrast to this official view, J. F. Kenney (2002) and other researchers of the *Russian Academy of Sciences* in Moscow, Russia, have produced inorganic petroleum in the laboratory. For thermodynamic reasons, they are convinced that complex hydrocarbon mixtures and thus petroleum in nature cannot come into existence out of strongly oxidized carbon compounds – which are dead animate beings. In contrast, it is possible to form spontaneously hydrocarbons (methane) arising from petroleum at high pressures and high temperatures. The suitable conditions prevail at greater depths than 100 kilometers in the mantle, but not in the near-surface sedimentary basins that were previously assumed to be origin of petroleum. The consideration of the Russian scientists have been confirmed experimentally. Petroleum is therefore a high-pressure type of methane, similar to graphite which can convert into a diamond at high pressure.

Contrary to this, in a laboratory petroleum could never have produced out of biological materials, even though huge funds were spent on this research. If one supposedly know how petroleum is originated, so should be able to reproduce the origin of petroleum in the laboratory. But this is just as difficult to replicate as the supposed start of life on the Earth's surface.

The researchers in the Western world will benefit from both unproven theories to a considerable extent, and it will be awarded many honors and doctorate degrees and research funds for the support of this thesis. But quantity does not replace quality. The theory about the origin of the biogenic hydrocarbons was in the U.S. and in Europe as such, of course, that was in a different direction and only partially researched. In contrast, mainly Russian researchers since the 19th Century represented the opinion that petroleum is abiogenic and not biologically caused. The famous Russian chemist Dmitri Ivanovich Mendeleev (also: Mendelejew), who account for the periodic table of chemical elements, wrote in 1877: "The most important fact is that petroleum was born in the depths of the Earth, and even there must we seek its origin."

In 1889, W. Sokoloff published a paper of the cosmic origin of bitumen, wherewith in the past was described the entire spectrum of oil to pitch and tar. He knew even then that some meteorites contain bituminous substances, and brought this fact in connection with bituminous substances in our soil. In this book, Sokoloff developed the hypothesis that the Earth was composed of meteorite-like substances and therefore originated as a low-temperature condensate, under conditions where hydrogen was in abundance.

He could find no correlation between the presence of fossils and deposits of hydrocarbons on Earth, and pointed to discoveries of petroleum and tar in crystalline bedrocks, such as, for example, in the basalt at the foot of Mount Etna. If hydrocarbons have been originated from biological remains, it should hardly to find petroleum in granites and basalts, at the most in the form of slightly infiltrations. If hydrocarbons originate biogenous, this should occur only where biological remains (fossils) are found up to very small amounts: in sedimentary rocks such as limestone, sandstone and shale. But petroleum can be found, for example, in plutonic rocks, such as the Cambrian gneiss like in the springs on the eastern shore of Lake Baikal, in the circular Siljan Ring in the gneiss of central Sweden and in the bedrock of the Russian Kola Peninsula.

The famous Russian geologist and petroleum expert Nikolai Kudryavtsev (1959) is often regarded as the father of the theory of abiogenic origin of petroleum. He realized that in the deep under each major deposit (accumulation) is petroleum available in all geological layers. This circumstance shows no dependence on the composition and the condition of the rock formations that make up these geological horizons. Petroleum may,

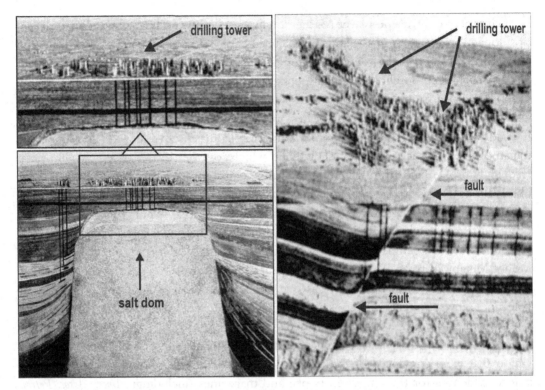

Figure 80: Petroleum occurrence's. Left picture: *A "floating up" salt dome, Petroleum bearing layers pushed upwards towards the surface. It was not until 25 years after the exploration of petroleum that was also discovered petroleum at the flanks of the salt dome.* Right picture: *Petroleum accumulation often extend along faults such as the Powell petroleum field in Texas, since gas and petroleum can flow more easily upward and those "deposits" (accumulations) regenerate faster. Images processed from Landes, 1970.*

depending on nature of the strata, sometimes be present in small amounts only, while commercially interesting deposits are located in permeable rock types, situated below a caprock. This system is called "Kudryavtsev rule", and it has been observed many confirmed examples (Kropotkin / Valyaev, 1984).

Thus, the *Lost Soldier* petroleum field in Wyoming has petroleum-bearing geological layers on every horizon of the geologic profile, which are overlaying the old crystalline basement. Petroleum is present in the upper deposits of the Cretaceous period as well as in the Cambrian sandstone, which lies directly above the crystalline basement. Northwest of Dallas in Texas in the gas fields I realized that there is to find generally gas down to the maximum drilling depth of about 3000 meters, but regardless petroleum is to find in each layer above the deepest layer (Barnett Shale) reached by drilling. The sediment layers positioned below these gas- and petroleum-bearing formation strata down to the Cambrian sand, which lies atop the supposed granite base have not yet been drilled, because such deep boreholes are still too expensive at present. But it should be possible to find hydrocarbons in the four layers which have not yet examined, because here extend a fault between two tectonic plates, where natural gas can easily migrate upward.

Kudryavtsev gave many examples of petroleum discoveries in crystalline or metamorphic bedrocks and in sediments which overlie those bases, which have frequently broken and from which the oil can stream upward easily. He adds a series of observations to his arguments, to which will be paid no attention anymore, because apparently nobody interviewed eyewitnesses. Thus, above violent volcanic eruptions occurred severe lightning's and during the eruptions of some volcanoes can be observed sometimes great columns of flames. One such example was lapping up 500 m at the Mount Merapi eruption in Sumatra in the year 1932. Kudryavtsev also pointed out that the loss of such a huge amount of gas that is released during violent eruptions of mud volcanoes, an subjacent gas deposit would have exhausted! Also, the fact that lava and mud volcanoes are often linked, he noticed

In the following years there lived various professionals and scientists in the Soviet Union that were convinced of the abiogenic origin of petroleum and gas and brought before evidences for this theory in books and magazines (including: *Amer. Assoc. Petrol. Geol. Bull*, vol. 58, 1974, p. 3—33).

But there were also such opinions outside the Soviet Union. Sir Robert Robinson, who was one of the most important chemists like Mendeleev, concluded due to appropriate investigation that petroleum contained much more hydrogen and was less oxidized in general than it would have been expected due to a biological origin. He was convinced of a hydrogen-rich primitive material that was *subsequent* contaminated with biological substances, because this matches closely the composition of petroleum. He examined the proportions of certain molecules in petroleum and found a remarkable agreement with similar proportions in synthetic petroleum. His studies were published twice in the science journal *Nature* (Robinson, 1963 and 1966).

The geologist Pyotr Nikolaevich Kropotkin took part in the search for petroleum in the western part of the Ural, and worked at the *Geological Institute of the Russian Academy of Sciences* (RAS). He received 1994 the *Demidov Prize*, one of the oldest and most prestigious awards for scientific achievement. In 1976 he came together with P. N. Valyaev to believe that petroleum is everywhere where the pressure conditions allow the condensation of heavy hydrocarbons, which are rapidly transported with an upward-stream from great depths. Kropotkin published in 1986, and clarified once again the theory of abiogenic origin of petroleum and natural gas.

According to Kropotkin and Valyaev (1984) "in the majority of accumulations can be demonstrated that a vertical migration of hydrocarbons are occurred out of layers which are far below formations with plenty of organic matter, which were themselves regarded as the starting material for petroleum."

Not only were presented many examples where the Kudryavtsev-rule is fulfilled, but also those described, where is to find petroleum and natural gas within or immediately above the (crystalline) bedrock and this in areas where exist no organic accumulations which could serve the explanation of the biogenic origin of these hydrocarbons.

In my book *Darwin's Mistake* it was explained that petroleum accumulations can be found in areas where meteorites smashed into the Earth's crust (Zillmer, 1998/2006, pp. 213). This observation was forced due to the study of the location of craters and petroleum fields. The connectivity is to see that there are *cracks* in the Earth's crust, where hydrocarbons are able to migrate over greater heights across very easily (Figure 80, right picture), *without* an step by step upward-cascade towards the surface (see p. 188 f.).

After some time, the famous astronomer, Professor Dr. Wolfgang Kundt asked me whether I have taken over this recognition from Professor Thomas Gold, but whose books I didn't know at that time. However, this led me to Gold's book *The Deep Hot Biosphere*, which was published in the same year as my first book. But already in 1992 he published some aspects of his theory with the same title in the journal *Proceedings of the National Academy of Sciences* (July 1, 1992, vol. 13, pp. 6045–6049).

Today Thomas Gold is considered as the modern representatives of the theory of abiogenic origin of petroleum, and he suspected, even with reference to the researches of Russian scientists, that large quantities of hydrocarbons are available in the depths since formation of the Earth and migrating upward continuously since this time. Consequently, almost inexhaustible natural gas and petroleum deposits exist in the depths of the Earth. What evidences are there for this theory in the Earth's crust?

Helium with Methane

As a noble gas, helium does not participate in any biological process. Therefore, a biological concentration of helium can be excluded. At first there were developed sensitive equipment's to detect helium, because it was believed that it is possible in this manner to find uranium deposits and these have been considered as sources of escaping helium. Although most of the helium on Earth results from the radioactive decay of uranium and thorium, this method was not very successful to detect uranium deposits. But it turned out that this method was successful to detect petroleum and gas fields.

It would be difficult to understand why helium – which do not originate biologically – share the same reservoirs of other gases which in contrast are believed to originate biologically. In many parts of the world, there is a close relationship between methane and helium. On its own this obvious fact casts doubt that methane could be of biological origin. Where does helium come from and why it is possible to find as much of this noble gas, more than all rock radioactivity's discharge? Since helium does not enter into chemical combination with another element, no such process can cause that helium passes from a lower into a higher concentration.

In many gas fields around the world occur helium concentrations of about one to three percent, in some cases up to ten percent in methane-nitrogen mixtures! All these high concentrations are found where methane, petroleum and often nitrogen is stored under an impermeable layer which is dense enough to prevent the leakage of helium.

"Because helium derives mainly from ongoing radioactive decay of uranium and thorium, it is only very diffusely distributed in the rocks. By itself, radioactively produced helium could not create pressures sufficient to open pore spaces in the rocks to allow any bulk flow to occur. Molecular diffusion of helium through the rocks would be the only mechanism for its ascent. Although molecular diffusion of helium would be faster than that of any other gas, it would nevertheless be much slower than bulk transportation. Helium transport therefore must be driven by another and much more abundant gas that provides its own motive force both for upward-streaming and for generating pressure-induced pore spaces and fissures along the way. It is the driving force of this other gas that provides the required pumping action that concentrates helium in hydrocarbon reservoirs near the surface. If that other gas is a hydrocarbon, it will of course pump the helium it has picked up into regions that we identify, at shallower levels, as hydrocarbon-bearing. This, then, would account for the association of hydrocarbons with helium.

The test of the hypothesis that hydrocarbons fluids serve as the upward carrier of helium would be this: If helium could flow without a carrier fluid, there should be many locations where amounts of helium had accumulated that were similar to the amounts of helium in some gas field; but in the absence of methane or nitrogen, these accumulations would be pure helium fields. Given the extent of geological exploration, many such fields should have been discovered by now: they would be of great value. Their absence thus supports the carrier-gas concept of helium transport" (Gold, 1999, pp. 75).

The relationship between helium and hydrocarbons is so tight-knit that the exploitation of helium takes place to a large extent only for commercial purposes on oil and gas fields. If helium could flow more than the maximum distance of a diffusion process without any carrier fluid, than should occur self-contained accumulations of helium which possess a helium concentration similar to gas fields. But no suchlike accumulations *without* methane or nitrogen were discovered until today. Ruther, at hundreds of monitoring stations on Earth was found in less depth – as that of agricultural water fountain – that there is a close relationship between methane and helium concentration (Welhan / Craig, 1983).

The deeper the source of hydrocarbons the more helium atoms result from the

Figure 81: Rising helium. *Helium is washed out of a little nitrogen in great depth and a lot of methane in higher areas.*

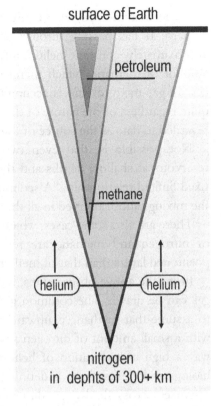

surface of Earth

petroleum

methane

helium   helium

nitrogen
in dephts of 300+ km

radioactive decay, and can be collected with upward-streaming fluids (gas). The helium concentration in the gas can therefore serve as an indication of the depth from which the gas had migrated upward. Corresponding calculations allow us a rough approximation and therewith to distinguish between possibly biogenic gases resulting from deeps down to about 8 kilometers and abiogenic from perhaps 150 kilometers depth. So we can also explain that the areal extent of the discharge areas of special gas mixtures is to be arranged in a relationship of the depth of the source. If gas mixtures escape in a wide geographical area, then we can assume a very deep-lying source.

Thomas Gold and Marshall Held analyzed data, which were collected from the U.S. Bureau of Mines from 1038 wells in oilfield areas in Texas and Kansas extending 320 kilometers long. They discovered that volatile of all elements and components of nitrogen (N2) and helium came frequently from deep layers, because some typical mixing ratios of these substances sometimes extend in the horizontal stratify near the surface over much larger areas than those in which methane was to find additionally. Because of these spatially limited distribution, natural gas (methane) has its origin in the next higher layer, and still higher occurs petroleum with various additives of other hydrocarbon gases. But all of these origins are lying significantly deeper than those sediments, which overlay the crystalline bedrock (Gold / Held, 1987).

Nevertheless, is it possible that methane originates in flat-lying sedimentary layers, while the helium-nitrogen mixture are upward-streaming from very deep wells and therefore penetrate the whole region? An origin of methane in sedimentary layers could not explain why the mixing ratio of helium-nitrogen gas fields possess specific and even not random values relative to the concentration of methane.

It seems to be the best explanation for the spatial distribution and according to certain patterns occurring concentrations of these substances that nitrogen is washing out helium, which is caused by radioactive decay in depths of about 300 kilometers. The such resulting helium-nitrogen mixture then had to cross methane-rich layers on the way upward. It has passing through this region and above escaping gas then exhibit an approximately constant helium-nitrogen ratio, but it would have collected a certain amount of methane, which therefore varies in diverse regions. Beneath the caprock a certain gas mixture ratio is accumulating without any influence of local circumstances - quite regardless of the nature of the geologic age and depth of formation, which has been regarded to date as the source of the gas.

Not possible is, that even two separate mixtures were made up of two *separate* reservoirs at shallow depths and therewith are able to provide all the drill-holes with a fixed limited relationship. "A sedimentary origin for the methane cannot not account for the mixing ratios observed in all these large commercial fields" (Gold, 1987, p 120).

There are also some cases, which are the other way around and where the appearance of nitrogen and methane are reversed. Then, the nitrogen source reside in higher positioned layers than that of methane.

From the above observations, a conclusion in favor of an abiogenic origin of natural gas can be drawn. The common global relationship between helium and methane allow to assume that methane is upward-migrating from deep layers and absorbs helium that with a small amount of nitrogen poured out of layers in even greater depths and in this way a high concentration of helium is delivered. If the mass of methane came from biological deposits, there is definitely no reason for the observed relationship. At short distances across flat-lying sediments, biogenic originated methane cannot collect more helium than carbon dioxide or water assimilate in long ways which are several hundred times commonly, but generally contain much less helium.

Carbonate Cement

The close relationship between helium and hydrocarbon deposits can only be explained satisfactorily by the abiogenic origin of petroleum and natural. Another strong argument is the fact to value that petroleum deposits regenerate over certain time periods. Moreover come along the characteristic of extensive occurring carbonate cements that fill crevices, but not occur as a geological layer.

It is found that in the carbonate cements have a wide range of carbon isotope ratios which can be attributed to an equally wide bandwidth in the upstreaming methane. Carbonate cements stem methane and therefore are common in oil-bearing strata. In addition, these carbonates cause a "cementing" of caprocks to make these less permeable, so that it is possible to build commercial deposits of petroleum and gas underneath.

The geological interest in carbon isotopes is well-grounded in the slight variations, resulting in case of different natural sources. The particular isotope mixture ratio tells us

Figure 82: Calcite dike. *Upright joints within compact limestone in the "Donnerkuhle" near Hohenlimburg in Germany are filled with pure white calcite (Kukuk, 1938, p. 35). Person (in dotted oval) for size comparison.*

something about the development of a certain carbonic material. On this basis, many oil geologists believe that the biological origin of petroleum and natural gas is proven. Let's look closer at this aspect ratio.

The natural carbon of the Earth contains predominantly carbon-12 and in a fraction of only one percent there exists carbon-13 as a stable isotope. There are no processes that would be able to fundamentally alter this seemingly pre-determined mixing ratio by the Sun. Only due to very specific processes, the ratio will be affected slightly in favor of either light or heavy isotope. This is called a process of fractionation. Small variations of the carbon-13 content of the sample compared to the norm – as the center of the range – are usually expressed as parts per mil, more or less of the norm.

The distribution of carbonate-isotope ratios in different forms of carbon is a explicitly indication of their origin. Marine carbonates, such as limestone and dolomite, laid down from atmospheric carbon dioxide dissolved from ocean water, range remarkably stable from about -5 to +5, as to see in Figure 83 (Schidlowski, 1975).

"Because the oceanic carbonate rocks got their carbon from this CO_2, the carbonate record should show a gradual increase in the proportion of C-13. Given the quantities estimated for the deposits and their isotopic ratios, this effect should be sufficient large to be observed in carbonates laid down over geological time.

But no such effect is seen. The carbonate deposits in fact show a small range of the isotopic ratio, which has stayed remarkably constant from early Archean times to the present. The biogenic theory fails to account this fact. This imbalance could not be redressed by the recycling of the unoxidized carbon deposits; these would largely turn into insoluble and heat-resistant element carbon.

On this bases of the abiogenic theory we would consider most of the unoxidized carbon deposits in the crust as derived from upwelling hydrocarbons, not from any sediments coming from the atmosphere" (Gold, 1999, p. 68).

As to see in Figure 83, petroleum shows a *very expanded range* from –20 to –38 per mill of isotopic ratios, whereas carbon in plants is relatively slighter and ranges from –10 to –35 per mill. It is just methane and calcite (carbonate) cements in the rocks which span a

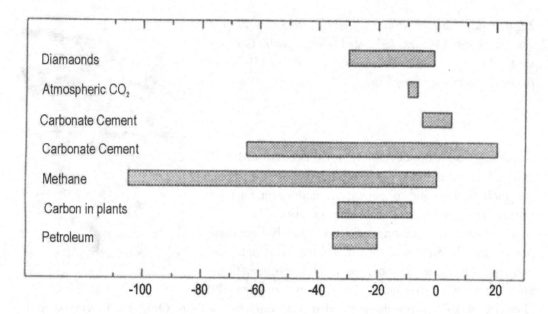

Figure 83: The distribution of the ratio of the stable isotopes carbon-13 and carbon-12 in various materials. *The wide distribution of carbon isotopic ratios in methane from all natural sources. Note that the range of atmospheric CO_2 is small, as it must be because all diverse contributions are quickly mixed globally in the atmosphere. The oceanic carbonates that derived their carbon from the atmospheric CO_2 show a similarly small range (Gold, 1999, p. 70). Methane and calcite span a much wider range of isotope ratios than all other forms of carbon on Earth. This close relationship shows: calcites are generally produced from methane.*

very much wider range from –110 to +20. This circumstance suggests that calcite in general be produced from methane. The displacement of the calcite to the heavier side indicates a fractionation, which occurs when methane is oxidized and then combines with calcium oxide to form calcite.

These calcites are to be found in areas of gas and oil field in large quantities, while smaller amounts may also be found elsewhere. With occurring of calcites and the filling in crevices, caprocks originated, where underneath may accumulate hydrocarbons to build rich deposits. The carbonate cements (calcites) occurring in many sediment layers are different from marine carbonates (Gold, 1999, p. 69):

• They fill in crevices and do not occur in layers.

• It is to find in oil-containing environment.

• They occur in a much wider range in the ratio of the two carbon isotopes from one another than with any other carbonates.

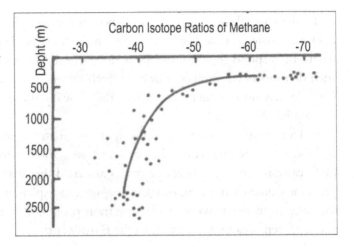

Figure 84: Carbon isotope ratios of methane plotted against depth of occurrence (from Galimov, 1969). Although there is much variability in this relationship, it is almost always true that where methane is to find at different levels in the same area, it is isotopically lighter (less carbon–13) the shallower the level. After Gold, 1987, p. 179.

It is interesting that "normal" clay – a common soil type besides sand and silt – emerge from calcite (feldspar), if carbon dioxide is added. If the clay is present in pure form and has a whitish color, than this is kaolin (china clay). A part of the clay deposits whose mass origin (conventional) seems puzzling, is due to the rise of methane, while above all clay minerals are originated at the surface due to weathering of other minerals.

Areas in which hydrocarbons occur, tend such to exhibit in all the layers underneath, right down to the base. This "Kudryavtsev rule" we can add the "Galimow rule," after which the methane tends to be isotopically lighter in each such vertical column, the higher the layer (Galimow, 1969). This is true for the large majority of the cases which were studied, regardless of the nature and age of the geological formation from which the sample was taken. On the way up, something is always lost due to methane oxidation. The resulting carbon dioxide readily reacts with the calcium oxide contained in the rocks and forms carbonate cement respectively calcite. This always reacts with the heaviest isotope. So the remaining methane is going to be isotopically lighter (less carbon-13) on the way up. In each geological layer, thus calcite emerge from fractionated methane.

This continuously, with the upstreaming ongoing fractionation is a very important process, because the each remaining material would be embossed in a much stronger fractionation than it is possible with a single chemical step. But the possibility that such a progressive fractionation and thus a reduction of carbon-13 could take place under natural circumstances was officially ignored, because there was considered no alternative to the theory of a biogenic origin of petroleum and natural gas. Although, the before discussed natural fractionation is not a stand-alone evidence of the abiogenic origin of hydrocarbons, so this scenario is an alternative theory to the commonly statement that the reduced carbon-13 concentration represents a clear evidence for a biogenic origin!

To deny the abiogenic origin of carbonates, it was claimed that abnormal, isotopically light carbonates can be found in areas where occur no hydrocarbon deposits. Directly it has to be respond that in these areas no gas and / or petroleum was found, but may be such deposits occur in deeper layers, which no one has yet drilled for hydrocarbons. The Geology has no possibility to predict theoretically the existence of gas or petroleum in the depths of the Earth.

In fact, there are also isotopically light carbonates in igneous and metamorphic rocks, and this occur exactly where the conventional geology had been denied the possibility of hydrocarbon deposits. Therefore it was assumed that carbonates would have a different, previously unknown origin. But just this, we can explain with the gas emission process of methane from deep layers. It is a common process that takes place regionally but with varying intensities on the way up to the Earth's surface.

No Petroleum in Arabia

The filled to the rim gas and oil deposits are not consistent with the geological observations. On the basis of currently taught geology has not fundamentally changed since 1920. So it's a beautiful story that for 15 years the concession for Kuwait was offered to the leading oil companies in the world. But they all declined, despite the prolific oil seeps in the area. All experts knew that there was no oil in Arabia, because it was geologically impossible (Pratt, 1952), and any documentation suggesting that the area was particularly favored with biological deposits was still missing. Finally, in 1937, the oil field that was discovered there proved to be by far the largest one known at this time. Even today, oil drilling will not proceed on the basis of geological knowledge. Therefore only one in seven boreholes is lucky to find petroleum, if it is not an add-on drilling in areas with producing wells. If seismic researches are carried out to explore, they do not know if subsurface there is water or another fluid instead of petroleum. Therefore, it is possible to pull up only salty water.

On the other hand is represented officially, that experts know how and where hydrocarbon deposits occur. Therefore *no* discussions take place anytime. The biological origin of petroleum and natural gas is a matter of certainty. Everyone knows exactly but no one can explain it. If a theory about the origin of the hydrocarbons will begin to obtain acceptance, one should be able to geologically explain such a vast oilfield in Arabia at least, if it was not possible to predict. What explanation is offered?

Conventional geology does not provide an answer, because the oil and gas fields of the Middle East extend over an area from southeastern Turkey to the Strait of Hormuz respectively Golf of Oman in very composite formations of different geological time periods can be found. This statement also applies to other areas with rich oil and gas deposits. There is no common feature that may explain why petroleum is concentrated in certain areas of the world. Looking more specifically at the oil fields in the Middle East, they have little in common. On the one hand, they are deposited in shallow sediments of the Arabian desert and on the other hand to find in the folded mountains of Iran. The gas fields and the overlying oil fields originated in totally different geological time periods and are covered by caprocks composed of very multi-support soil bodies.

Organic sediments, which could be clearly viewed as a source for the largest oil fields in the world, were not definite decided and are therefore disputed by experts. The reason can be seen in the fact that the quantity of organic sediments shall be deemed to be insufficient for the huge oil and gas resources. This applies especially to the fact that there exist extensive natural seepages in this area (Barker / Dickey, 1984).

Geological investigations confirm: "It is a remarkable fact that the richest oil region of the world has leading to a lack of conventional mantle rock." It is further stated that the oil-bearing reservoirs of age from middle Jurassic Period (about 164 million years ago) to the Miocene (5 million years ago) are distributed with a maximum in the middle Cretaceous and the Oligocene. It is confirmed, that nevertheless there is a "remarkable homogeneity in the chemical composition of the oils, and there is a presumption that they have a common stratigraphic origin" (Kent / Warman, 1972).

It was also noted, that coal deposits were found in the mountains of Persia as well as in the plains of Arabia in addition to gas and oil deposits. Is it possible, that the coal origination be associated with those of gas and oil?

Should be offered an explanation in terms of a biological origin of only a single one of the many oil fields, this is insufficient as long as the general plentifulness of the region and the common nature of all the oils cannot be explained. For these reasons, the usual reference to a "rich biology" does not settle the biology origin of hydrocarbons once and for all, because this statement cannot apply for the entire area in its expansion from southeastern Turkey to the output of the Persian Gulf and this in particular for time reasons, because these formations are spanning geologically time periods of more than 160 million years. But it is believed, that huge amounts of sediments containing organic matter could have been buried by caprocks to create the unusual abundance of "fossil" deposits.

One have to ask, where did come from all the material for the impermeable caprocks, which differ in their chemical composition and varies in terms of the geological age. How fast did this process occur? Imperceptibly slowly to the geology underlying theory of uniformity (Lyell's hypothesis)? Biological relics should have been buried in this region over a time span of 160 million years, and at any one time the geological formation respectively deposits must have been covered with caprocks very slowly, little grain of little grain? Without a rapid covering no biogenic oil field can develop! Therefore, it is an unlikely scenario that would have validity for all other oil regions. Required are numerous fortuities in abundance, and this throughout the world!

Because the composition of the Earth's mantle and crust in this oil region is extremely different, another explanation should be found. Let's assume for the purposes of the abiogenic origin, that an area of the mantle is very rich in hydrocarbons. As a result, fills and fills to overflowing every as a pocket-acting "trap" in the overlying geological layer. Surplus crude oil rises to the surface. This happens regardless of whether these reservoirs suitable for storage are located in flat plains or on steep hills or in deep or shallower layers and this regardless of the geological formation age. If the source of the hydrocarbons is to find deep below, then chemically closely related raw oils are necessarily able to penetrate into areas of the crust from below. Therefore the petroleum deposits differ by local features in any respect and it is not surprising that it is determined by investigations that the most reservoirs are filled about to overflow level (Kent / Warman, 1972), and the world's oil reserves are bigger than ever; also in the universe? In November 2012 in the horsehead nebula Orion was discovered petroleum.

In contrast, due to the almost unimaginable plentifulness of hydrocarbons in the Earth's crust, one will explain this fact in terms of the biogenic theory and therewith with the existence of a powerful carbon dioxide shell of the primordial atmosphere, then this shell must have been amazingly mightful. From the bulk of currently known carbonate rocks it is possible to suggest a mass of carbon dioxide, which was about eight times larger than the total mass of the present atmosphere, and about as dense as the atmosphere of our sister planet Venus. There are good reasons to believe that the Earth has possessed originally not so much gaseous material. One of this reasons is the low percentage of inert gases such as neon, argon (if it was not produced during nuclear fission), krypton and xenon in the atmosphere today. No physical process could have taken out this noble gases from the original gas mixture of the solar system, in which – as we all know – are abundantly represented accordingly (Gold, 1999, pp. 62).

In addition, there exist an important aspect in terms of the age respectively the geological periods in which the carbonate rocks shall have originated.

Instead a slowdown of the carbonate rock formation rate over time due to the decreasing carbon dioxide in the atmosphere it is rather to recognize a fairly steady increase of oxidized as well as unoxidized carbon and this over a time period of the last two billion years – that's the time of sedimentations at all. It even shows very clearly that the amount of carbons in the surface layers of the Earth has *increased* over geological time periods. The propagated closed carbon cycle of biogenic origin cannot serve it. Instead of this we must have a steady outgassing from sources deep in the Earth (ibid., p. 83).

Where did all the carbon dioxide come from, that kept at a certain amount in the atmosphere and the oceans, so that carbonate rocks were deposited over long geological periods of time and nevertheless the carbon concentration in the air was high enough, that the plants could live on it? Subject to these conditions the average rate of outgassing

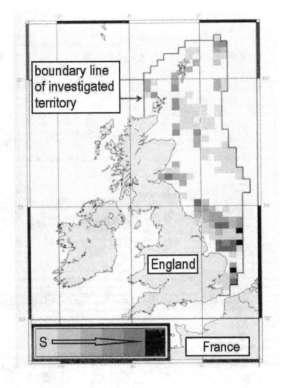

Figure 85: Methane seepages.
Strength (S) of the suspected density of gas seepages in the North Sea sector of the UK. Edited from Judd et al., 1997.

would have replaced the stock of carbon dioxide in the atmosphere and oceans approximately every 2.7 million years. The outcome of this is, that the supply of carbons would have been 740-times renewed in the last two billion years.

Because the biogenic theory empirically cannot confirm the described situation and scenarios, it was not surprising that an adviser to the British government announced prior to drilling, he would drink every cup of oil, one would get from the North Sea. Like the ruling predictions and opinions he believed that this is a hopeless area to drill for petroleum. In fact, there are huge gas and oil reserves as far north as Spitsbergen. In today's ice-covered areas are to find not only huge coal deposits, but also large quantities of methane (hydrocarbons).

There are many gas and oil fields, which were found by chance and for whose existence there is not sufficient geological reasons. But let us go to an area again where the terrible tsunami occurred in the Indian Ocean in 2004. It was found that the East Pacific Rise, a rift valley in the Pacific Ocean, emits methane along a large part of its length, together with very hot water, why it is not possible there to form methane hydrate (Kim et al., 1983).

In principle, the East Pacific Rise is a crack in the Earth's crust, where two tectonic plates are separated and where the very thin seabed millimeter by millimeter shall slide under the thicker and much heavier continental plate. With this process of subduction shall be forced an "elastic" storage of stresses and in case of a stress relaxation can be triggered a tsunami. This is technically a not to define proof or a predictable process, what therefore constitutes an explanation solely owing to the plate tectonic theory. But this seemingly too low energy pulse to trigger a tsunami could be shown multiple times in the German television news as the sole reason and was therefore manifested as a "fact".

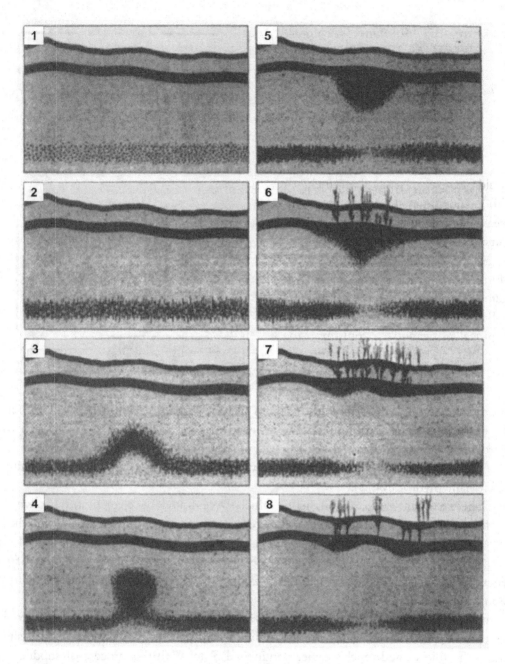

Figure 86: Bulging gases as the cause of earthquakes. *Fluids, including hydrocarbons from a depth of perhaps 150 kilometers, cause the rock brittleness and breakage. The strength of the rock is reduced until it breaks. earthquakes are triggered. Their cause is not an increase in the internal stress of the rock, but conversely, a "fatigue" of material and a reduce of the break strength. Such may explain why aftershocks usually seize a larger area. After Gold / Soter, 1980.*

The East Pacific Rim is the best-documented example of a region that has no extensive sediment layers and from which is still methane outgassing, along with helium and carbon dioxide also in great abundance. Essentially, this is a volcanic area, but along this crack (= trench) in the island arc of Indonesia also are to find large mud volcanoes (Fig. 19). According to our interpretation, all the terms of the theory of abiogenic origin of hydrocarbons are fulfilled in this area. The rising carbon dioxide here is the oxidation product of methane, which can only upward-migrating from the depths of the Earth, because there are no thick sediment layers on the ocean floor, which can be regarded as the sources of gas. At the same time many earthquakes occur, which are generated by rising hydrocarbons. The source mechanism is also responsible for the activity of the lava and mud volcanoes in this area.

Also other island arcs, like the East Pacific Rim, show a similar relationship between volcanoes, earthquakes and the occurrence of gas and oil in considerable quantity. The Kuril Islands, an approximately 1,200 kilometers long island arc with more than 30 islands which connects the Russian Kamchatka Peninsula and the Japanese island of Hokkaido like a bridge, possess about 100 volcanoes, of which 39 are active. Earthquakes often occur, and there are remarkable hydrocarbon seepages. It was estimated how much methane would be risen to the surface in the 80 million years of its existence (according to the geological time scale), if the current amount is assumed to be constant. It is around 380 trillion (= 3.8 times 10^{11}) cubic meters of gas, which is vastly more than the content of any knows gas-producing area of similar dimensions (Kravtsov, 1975).

The just described two chains of islands show the same pattern as other island arcs, "namely an ocean trench to the outside, then a line of active volcanoes as one moves inwards across the arc, a deepening of the sources of earthquakes a line of hydrocarbons occurrence, mainly paralleling the active volcanoes, but to the inside of the arc". We can expand this system, because we have already discussed the global model, "made up by the lines along which earthquakes are common, and the strong relationship these show to commercial hydrocarbon occurrences. Of course there are plenty of hydrocarbons occurrences outside the seismically active areas also, but nevertheless it is clear that there is a strong positive correlation" (Gold, 1987, p. 131).

Not only alongside island chains, but there is a general relationship between active volcanoes and gas and oil occurrences, because pipes combine deep layers with such near the surface. Methane migrates upward and is oxidized, originating carbon dioxide and water with release of heat, while non-oxidized hydrocarbons accumulate at the fold flank of volcanoes. So gas and oil was extracted at the fold flank of Mount Etna in Sicily, and it is reported that over volcano crater appeared flames (Hope, 1824, p 239). The Mountain Oil (petroleum), which often appeared in Sicily, was to find on water tanks in the village Petralie (which got its name from), on water sources at Mistretto, Lionforte, Ivona; those from the area of Agrigento already reported Dioscorides and Pliny in ancient times.

Bitumen and asphalt occurred also and of sulfur Sicily was very rich (ibid., p. 250).

Also Charles Hoffe reported on the phenomenon of gas blowing outs in some places in Sicily in 1824. On countless small mounds of gray chalk, the leaking gas formed a hole in the center of the mound. Nearby Caltanisetta not far from Terrapilata, some of these hills got cracks, if earthquakes occur far away from them (ibid., p. 248).

In such a lava-volcano area with gas leakage and violent earthquakes, which are accompanied by deep rumbling over a period of time, there were also to find mud volcanoes in the form of soil liquefaction. Such an extraordinary phenomenon occurred on 18 March 1790 on a high plane, a few miles off the southern coast, where Terranova is located. At first, in the village was to hear a subsurface strong roar. The day after a quake, the ground deepened three Italian miles in circumference gradually, and at one point up to thirty feet deep down. Then into the subsided surface fractured an opening of about 30 feet in diameter, through which three hours a stream of mud came out with great violence. The mud was salty, it smelled "like sulfur and petroleum" (ibid., 249).

Already on 29 September 1777 when one of the largest eruptions of Mount Etna occurred, was heard for miles around a flannelly roar and on several Italian miles shook the ground, and in the middle of the area where opened a large deep hollow, rose a mighty pillar of mud mixed with stones of various sizes to a height of about 100 feet. The explosion took half an hour, then it was quiet, but after a few minutes it exploded again, and this phenomenon was repeated several times throughout the day. At the same time spread far around a strong smell of hydrogen-sulfide gas (ibid, pp. 246).

These eye witnesses describe accurately the side effects described in this book, if methane migrates upward. The conventional geophysical system provides a plausible explanation for the occurrence of earthquakes *within* tectonic plates nor for the rupture of mud with stones and escaping of hydrocarbons. Allegedly, the volcanic heat shall helps to dry organic sediments and convert it into hydrocarbons. But the amounts of sediment in Sicily are very little and the natural rate of volatilization of gases in the neighborhood was immense. However, such regions are not providential for the production and retention of substantial amounts of methane and oil.

Calabria, southern Italy, the south-western peninsula and the eastern tip of Sicily up to the volcanic shield of Mount Etna consist of granite, so that the great volcano stands at the edge of the primary strata. Because the granite base is apparently an impermeable caprock, hydrocarbons are able to swelling up only at whose edges. The islands Stromboli and Volcano are located north of Sicily located volcanoes – both belonging to the Aeolian Islands. These islands seem to be in contact with Mount Etna. Thus, in 1963, eruptions and movements of the two volcanoes signalized eruptions of Mount Etna, and earthquakes in the area around. It is interesting that Vulcano burned constantly, but at the same time Stromboli had showed only scare flames (ibid., 260). There is a clear indication that gas discharges took place.

Geophysicists are baffled, on the one hand because deep earthquakes which occur in subduction zones met in opposition to the theory of geophysics (Frisch / Meschede, 2005, p 115) and on the other hand, why they even not occurred at the edge of a tectonic plate but instead of this in the middle of tectonic plates like in the North of Sicily. Also unexplained are the stockwerk-like path of deep earthquake hypocenters upward in direction to the surface. In the past it was frequently referred to such events. The crash of ceilings, that are caprocks which overstretch cavities, happen not at the same time, but one after another, so that in such a manner described kind of deep earthquakes is associated with the process of upstreaming hydrocarbons and the occurrence of lava, mud, petroleum and natural gas – as the mode of action is already described in this book. Because the abiogenic origin of hydrocarbons is ignored officially, geologists are not able to predict where to find deposits of petroleum and natural gas. The geological-geophysical system is wrong! After all, if one would have assumed that deep down in the Earth, there a biogenic developed source would produce *all* amounts of hydrocarbons, then, consequently, the individual layers, which are *each* assigned to a particular geologic age, must have absorbed carbon at *different* geological time periods. Therefore, geologists should have to give an explanation for a large, 6000 meter deep gas field in the Persian Gulf, still existing under the abundant oil fields of Abu Dhabi which occur in shallow depths. Since, according to the conventional view, it is considered that there is no single source of hydrocarbons, then in terms of the biogenic theory must be accepted that this petroleum "at widely separated times has repeatedly formed in the same place" (Stutzer, 1931, p. 328) – *an adventurous scenario*!

As an example we can point out the deposits in North America and also in Wietze near Hannover in Germany. There are several overlying geologic horizons. One should be the Late Triassic (Keuper) which is believed to more than 200 million years old, while the sixth overlying oil-bearing layer is expected for the Late Cretaceous with an age of more than 65 million years by means of the geological time scale (cf. Kraiss, 1916). "Favorable conditions for the emergence of petroleum appear to withstand in northern Germany *at different geological periods at the same place*" (Stutzer, 1931, p 328).

This means, according to the biogenic theory of petroleum genesis, that dead marine organisms were deposited and in such a way accumulations were originated one about each other. In each case this must had happened over many millions of years and all deposits came into being in the same location. Every deposit must have been covered by a dense top layer (caprock), so to build separate geological horizons, in which dead marine organisms was transformed into crude oil at any one time. This also means that there will have arisen the same transformation process at every geological horizon in case of low temperatures and pressures. A typical conventional approach, due to the basic principle of uniformity (actualism), which indeed contradict given facts and findings, but nevertheless exhibit the basis of our worldview in general.

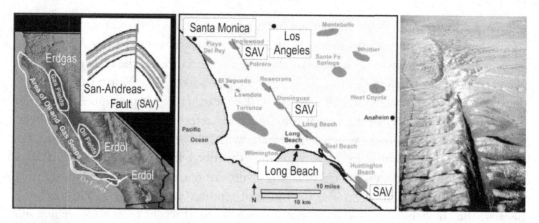

Figure 87: Fracture zones. *Left photo: Along of faults, such as the San Andreas Fault (SAV) in California, are seeping hydrocarbons and it is possible to haul oil and gas. Middle image: A section of the map of Los Angeles area with prospective oil and gas areas along and parallel to the fault zone, which is a 1,100 km long fault line due to vertical and longitudinal displacement in the Earth's crust (right picture). See Fig. 80.*

Therefore the assumption of "primary bedding, thus the replication of oil formation in the same place at different times" was modified in the German text book *Crude Oil* by Professor Dr. Otto Stutzer: "In almost every oil field there exist several oil horizons above the other. An oil field with a single oil horizon is a rarity. As exhausted believed petroleum deposits sometimes began to sputter again, if you drilled deeper and discovered new oil horizons" (Stutzer, 1931, p. 326). *Precisely this statement is correct.* The refill of petroleum deposits is a certain widely unknown fact, but indeed this scenario we had already discussed as an evidence of the abiogenic origin of hydrocarbons, because natural gas and petroleum are upwelling in the form of a stockwerk-like cascade if the break strength of the overarching cap rock is exceeded.

If we suppose that the sources of the hydrocarbons cannot exist in sediment layers, which are generally sat above the crystalline basement, but instead of this reside much more deeper, then the vertical horizon-like stacking would be expected to be a common feature of the distribution of oil and gas. Therefore it is no mystery that in the Hugoton fields of Kansas one can pursue this vertical stacking not only down to the basement, but actually into it. In that area, as perhaps in many others, it is possible to recognize the relationship between producing wells to the fault patterns underneath (Figure 88).

If oil should seep into the opposite direction – which means from above into the underlying crystalline basement rock – the amount of those downward-streaming fluids should be very small. Landes wrotes: *In fact, a commercial oil production takes*

214

Figure 88: Tube-like oil deposits. *An example of tubular arranged oil deposits: The "Garnett Shoe String" in eastern Kansas. According to Rich, 1923.*

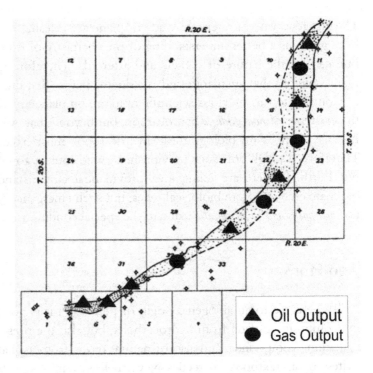

Place at numerous drillings in the basement rocks of the Central Kansas Uplift, which is believed to be 590 million years old (geological age: Precambrian). The first discovered well produced daily 1800 barrels daily from fractured quartzite. Subsequently more than 50 producing wells have been completed in the basement rock in this area (Landes, 1970). But normally the basement rock occur in depths which are too deep to reach with actual drillings.

But as already stated, there in Kansas, like in many areas as at the Llano Uplift in Central Texas, it is possible to recognized a relationship between producing wells and the fault patterns underneath. If are to find a shallower well with upward-streaming hydrocarbons above of fractured basement rocks then there is a high possibility that exist further wells along the fracturing pattern (Figure 88). Petroleum prospectors have known this for a long time, but it is not clear what explanations they had for it. In our view a deep-seated feature known to allow the upwelling of hydrocarbons in one location is likely to do so in other locations along its lengths of a rock anomaly (Gold, 1987, p. 133). The before discussed facts confirm the abiogenic of hydrocarbon:

• The fill-up of expended petroleum deposits
• The isotopic composition of oil grades
• Other characteristics of the extensive carbonate deposits in the Earth's crust
• The close relationship between helium and hydrocarbon deposits
• The horizon-like stacking of hydrocarbons
• The upwelling of methane from the depths through migration and diffusion.

215

The last few years of intensive study of planetary celestial bodies by many space probes have given us a better understanding of the chemistry of other planets and their moons, but also for the nature of comets and asteroids. Therefore it is at the time to examine many opinions that are manifested as a dogma before that time of space travel.

Contrary to this previous scientific opinion, on planetary celestial bodies were proved the existence of *non-oxidized* hydrocarbon; but beyond that even now in *very large quantities*. Because there is no biology these hydrocarbons must have been an abiogenic origin. Therefore, it's high time such previously negated sources to take into account to occur in our Earth. If there are ruling ideal-providential circumstances, it may be possible to originate crude oil from biological relics, but such emergence (genesis) will remain limited to *few deposits and small quantities* due to the special conditions.

Too Hot?

"If we agree to an abiogenic origin of petroleum it is possible to explain ... not only the composition of oil from hydrocarbons, but also the presence of nitrogen and sulfur containing compounds, and also of natural resins from asphalts," Stutzer (1931, p. 299) writes in his textbook correctly. Nevertheless, until today two main arguments were brought forward against the abiogenic (inorganic) origin of petroleum. So petroleum is contaminated with remains of biological molecules, which are found therein. In fact these molecules originate from membranes of dead cells. We will discuss this serious objection against the abiogenic theory after discussing of another major objection.

In consensus with the vast majority of scientists Otto Stutzer offers a purely chemical reason against an abiogenic origin at great depth, since the condensation points of most components of crude oil are between 0 and 300 degrees Celsius. "The crude oil would have upwelling in the form of vapor from deeper hot zones. One would expect that only certain parts would have condensed at different levels depending on the cooling-down of the ruling temperature; more in depth the heavier component parts, more above ones (ibid., 300).

Actually, the heat resistance of hydrocarbon in the hot depths of the Earth will be central problem. If the theory from a formerly hot Earth is correct, then first of all the volatile components would be ascent to the surface. After cooling-down of the Earth a further afflux of hydrocarbons from the deep would be impossible. Therefore one would expect an outgassing process, so long as the temperatures rise in any part of the Earth. Accordingly, an initially hot Earth had been brought most fluids to an energetically low chemical state and this already at an early stage. Therefore, reactions with a corresponding energy release could not occur later on, and all substances were in chemical equilibrium. For the start of life then would only remain sunlight as an energy source and microbes would be able to develop only at the surface. This is the basis concept of our conventionally view of the world – since starting of sound science.

However, if we underlie a *cold and even not hot* Earth at the beginning, then the volatile components are often not in chemical equilibrium with their environment and therefore could be a source of chemical energy. This viewpoint was already represented by several researchers, including Pascual Jordan (1966) and Thomas Gold (1985). For the energy problem and even more to explain the origin of life is therefore to understand and to explain the oxidation state of carbon within the Earth.

In this book are presented various evidences and advices to expose that the Earth was primarily cold and not hot. Nowadays our planet is only partially melted and this occur only in the upper mantle – in my opinion, mainly in the *neutral spherical shell*. This opinion has recently been confirmed as the source of the magma is not detected at very great depths in the mantle, but mostly in a chamber near the surface, as it is confirmed now in case of the Yellowstone volcanism (Bindemann, 2006, pp. 43 f).

The abiogenic theory of hydrocarbon originating depends on the key-question, if it is possible that hydrocarbons are thermodynamically stable at *great* depths. The fact is that temperatures above 600 degrees Celsius would dissociate methane – even the simplest and most heat-resistant hydrocarbon – and at the same time temperatures as low as 300 degrees Celsius are sufficient to destroy most of the heavy hydrocarbons components of crude oil. Because such temperatures are reached at depths of only a few of kilometers in the crust, scientists felt confident that hydrocarbon cannot survive at *great* depths; least of all not deeper than sediment layers exist. Actually it is true, if we take in account this temperature range *solely* it is not necessary "to discuss an origin of hydrocarbons from non-biological sources at deeper levels. If the origin had to be found in the upper and cooler parts of the crust, then there was really no alternative to the biogenetic theory" (Gold, 1999, p. 50).

But we must assess the question of hydrocarbon stability not only at the temperatures, but also at the pressures, that prevail at various depths. But calculations of thermal stability that were undertaken in the West did not take into account the substantial effects of pressure. So scientists saved very expensive high-pressure and high-temperature experiments. But we know that high pressure greatly stabilizes hydrocarbons against thermal dissociation.

Thermodynamic calculations made by the Russian scientist Emmanuil Chekaliuk show that the chemical decomposition of methane not take place to a depth of about 300 kilometers, as long as the temperature does not exceed 2000 degrees Celsius. This is the case in volcanic regions, because in these the temperature profile is different (Chekaliuk, 1976).

"Perhaps a depth of somewhere around 600 kilometers would be the lower limit for the possible existence of methane within the Earth." At a depth of, may be,

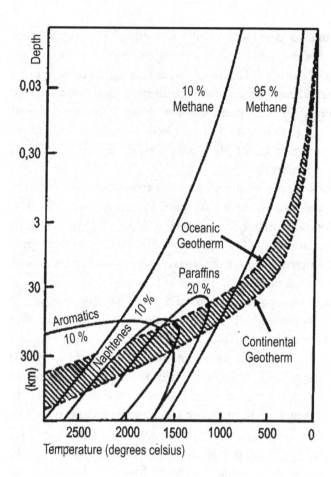

Figure 89: Hydrocarbon resistance (after Chehaliuk, 1976). The increase of temperature with depth in the Earth is referred to as the "geotherm", and the region between the two geotherms is the region (shaded pictured) that might represent the temperature-depth relation in different locations. The deep ground under the oceans is generally hotter than the deep ground at the same depth on the continents, as can be seen in the figure. Thermodynamic calculations indicate the domain in which various hydrocarbon molecules are stable. Methane is the most stable molecule of the hydrocarbons; most of it survive at all levels down to 300 kilometers, provided the temperature there did not exceed 2000 degrees Celsius. For the other components of natural petroleum – paraffins, aromatics, and naphthenes – the percentage in equilibrium are shown; these would be the values most likely to be produced from a mixture of hydrocarbon and carbon. Most of the petroleum components would be present in equilibrium at a depth between 100 and 300 km. Methane upward-streaming from great depth could bring up, in solution, significant fractions of these petroleum components towards the surface (Gold, 1987, p. 137).

"200 kilometers, a mix of hydrocarbon molecules would be the expected equilibrium configuration. The detailed chemistry of the resulting molecules would depend on pressure, temperature, and the carbon–hydrogen ratio. Other atoms that might also be present, such as oxygen and nitrogen, would form a variety of complex molecules with the carbon and hydrogen" (Gold, 1999, pp. 50).

Petroleum has already been found in significant quantities below 5,000 meters. In these cases, producing bitumen seem not to have destroyed up to temperatures of 300 degrees Celsius (Price, 1982).

Thermodynamic calculations done in Russia and in the Ukraine have suggested not only that most of the heavier hydrocarbon molecules, that make up the bulk of petroleum, are stable in the pressure-temperature regimes that prevail at depths between 30 and 300 Kilometers (see Figure 89) but also that they would be generated if a mix of simple carbon and hydrogen atoms were present at those depths (Gold 1999, p. 50).

In what mode mixtures of hydrocarbons exist in such depths? Even though methane is technically a gas in all circumstances under discussion, it will behave chemically like a fluid at high pressure. Such pressurized methane will be a good organic solvent, and all the liquid hydrocarbons and many of the solid ones are soluble in it. The solution, if the proportion of methane is high, will be a low-viscosity liquid, that can penetrate very through existing cracks and fractures in the rock.

High-pressure gases are present in liquid densities, one called this also "super–critical" gases. Substances, which are upstreaming with such fluids, precipitate in cases of a *sudden* pressure drop, a process which is also used in industrial applications. But this are precisely these sudden changes in pressure, we had already discussed in the context of the stockwerk-like outgassing process. This means that at each outgassing process into the next stockwerk some of the dissolved substances precipitated, or in other words, are thus deposited. So we can explain the layer-wise storage of hydrocarbons.

"All of these deeper pressure levels are below the range that we reached by drilling until now. But we may see some such deposits in rocks that have been uplifted and thereby made accessible" (see Figure 80, p. 197). "Concentrated deposits of carbon are known in many ancient rocks and may have resulted from this process.. The domain which we know well, however, is the one above the last major pressure drop that an upstreaming gas would suffer. It is the domain where gas reaches the hydrostatic or normal pressure, with which the petroleum industry is familiar. In sediments this last pressure drop for the gases (or first sudden pressure increase for the drillers) is generally between 3 and 6 kilometers depth" (Gold, 1987, p. 140).

After upward-streaming from the basement rock, the hydrocarbons lose hydrogen as a result of an ongoing oxidation. Once a hydrocarbon molecule coincides with a footloose or also by a catalyst (microbes) submitted oxygen atom, it captures this and loses simultaneously a carbon dioxide atom and two hydrogen atoms, from which may then arises water in a geological stockwerk-like layer – called *formation water* – with this oxidation process.

Of course, other atoms such as nitrogen or sulfur can take the place of the hydrogen atom instead of the oxygen atom, if present.

In this pursuit of reaching a chemical equilibrium it is to see the reason that hydrocarbons are going to be oxidized with the upstreaming process, whereas is connected a gradually loss of hydrogen. Therefore increases the ratio of carbon to hydrogen and many oil fields are layered like a pyramid cake. At the bottom there are a lot of methane deposits which are overlaid of slighter oils, whereas on which the heavier oils with much carbon and less hydrogen are deposited. The heaviest oils are on top, although above each layer may accumulate a certain amount of methane. But in some deposits the highest layer not contain the carbon-richest oils to complete the "cake layers" but still on top it can be deposited black coal The larger the carbon content, the more hydrogen was released and the more water as well as dioxide could be formed. This allowed a direct forming of water-bodies on black coal layers. In such sees directly deposited on black coal seams (= carbon layer) dinosaurs lived and swam, left footprints on top of the carbon layer, because this black coal was squashy at this time (see page 10).

Therefore not seldom black coal seams lie above oil deposits, such as the San Juan Basin in New Mexico or in the Anadarko Basin of Oklahoma. As to find in Alaska or Saudi Arabia. Almost, all countries that are known for their large oil reserves also have large coal fields. Also it is a common feature that coal is on top and oil below that chance cannot possible account for it. In Wyoming for example, some coal is actually found *within* the oil reservoirs.

It was shown that in zones with upwardly flowing hydrocarbons there is a strong tendency to form deposits of carbon, which grow into large concentrations. As a normal case, a fractionation take place from bottom to top, on the one hand to result in an accumulating of carbon and on the other hand to form (abiogenic) water and carbon dioxide with an oxidation process if oxygen is present. However, oxidation is not the solely reason why hydrocarbons lose hydrogen on their stockwerk-like upwelling. At high pressure, in the mantle created complex compounds are unstable with decreasing pressure near the surface, and hydrogen atoms gradually break off.

Coal on Top and Oil Below

The fact that coal is to find on top and oil below but also the geographical distribution of coal deposits poses a serious problem which is not possible to explain with the conventional theory. It is preconditioned that oil and coal are the result of completely different types of biological deposits which were laid down in quite different circumstances and where both occur at quite different geological times. Biological debris from marine algae is usually invoked for the petroleum genesis, and terrestrial vegetation for coal. No close relationship between the geographical distribution of the two substances would be expected. But in fact, as the oil and coal maps of the world have been drawn in ever-increasing detail, a close relationship has become unmistakable. The coal and oil maps of southeastern Brazil are striking in this respect (Figure 90). Indonesia presents another such example; local lore among those who drilled there for oil was: Once we hit coal, we knew we were going to hit oil (Gold, 1999, p. 97).

Figure 90: Problems for the biogenic theory. Overlap in the distribution of coal and oil deposits in eastern Brazil as one example of many other areas of overlap. After gold, 1999, page 98. Oil map adapted from International Petroleum Encyclopaedia, 1994, p. 95; coal map adapted from a commercial atlas by H. M. Goushu Company, San Jose, California (drawing adapted from Gold, 1999, p. 98).

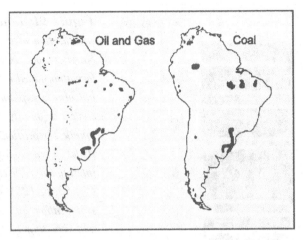

Analogous to the biogenic theory a *coalification* called process should occur. Fresh plant material shall be converted one after another into brown coal, bituminous coal, anthracite (carbon content of about 91%) and at least into graphite respectively stable carbon in some cases. As scientists admit, it was not possible until today to understand this process as part of the biogenic theory completely. But, the abiogenic theory can explain the origin of carbon deposits?

Because we recognized as a basic principle, a process of increasing carbon-concentration from the depths of the Earth in upward direction to the surface in particular geological horizons, this shows that peat and brown coal do not fit in the systematics of the inorganic respectively abiogenic origin of hydrocarbons and deposition of carbon, by reason that these layers *contain too little carbon*. In fact, in peat and brown coal, the structures of the original plants are still clearly visible. This is definitely cause the products of the Earth's surface biosphere. In the contrary, bituminous coal and anthracite fit in the abiogenic classification of vertical horizon-like stacking from the bottom to the top with increasing concentration of carbon, in particular because black coal deposits are often overlay oil and gas resources, which can hardly be a happenstance.

Our assumption is supported by the fact that nowhere intermediate stages of a gradual transition from peat and lignite to bituminous coal was discovered. Always is reflected a sharp borderline between black and brown coal, and by the upward-movement of coal-bearing strata in the area of contact can occur a so-called contact metamorphism. But no one will be found a gradual transition from brown to black coal, as well as no transitional forms of large animal (macro-evolution). Therefore the coalification process should have a different classification, analogous to the macro-evolution.

Against the biogenic origin of petroleum and coal is the fact that no one has ever put some biological materials in a container to concoct for suchlike "fossil" fuels. In contrast,

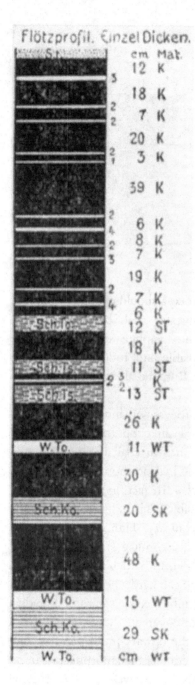

Flötzprofil. Einzel Dicken.

cm	Mat.
12	K
18	K
7	K
20	K
3	K
39	K
6	K
8	K
7	K
19	K
7	K
6	K
12	ST
18	K
11	ST
	K
13	ST
26	K
11.	WT
30	K
20	SK
48	K
15	WT
29	SK
cm	wT

Figure 91: Fine stratification. An example of the rich stratification of a thick single coal stratum from the Ostrava-Karvina coal basin. The conventional bog theory of coal formation failed not only with regard to such fine stratifications, but also to explain the origin of foliated coal, the aquifuge clay shale or mostly only little finger thin coal seams, often in countless repetitions with the same filmy middle shifts of clay shale. Thin coal seams of only a few centimeters thick intercalated with equally thin middle shifts of dirt bands (see Fig. 53, p. 128). How then could preserve the well-defined fine stratification and the large areal extent of such middle shifts and very thin coal stratums till this day during millennia of bulging and land subsidence of the (Lyell-)soil and this in or rather between several raised bogs according as to Potonié?
K = coal, SK = foliate coal, ST = black clay-sandstone, WT = white clay.

a laboratory produced petroleum pure chemically respectively inorganically (Kenney, 2002).

Just as it had existed relatively far too little microorganisms to make possible a biogenetic origin of the whole oil, so there were growing not enough quantities of plants and trees, which could rotted in swamps to convert into coal respectively to form all coal deposits. The *continental raised bog theory* of the German botanist Henry Potonié (1905) breaks down due to for lack of sufficiently biological mass. If we imagine a 100 year old beech grove, this would deliver building material of only two centimeters respectively as less as *an inch thick* coal stratum. For such, on the spot of the bog (indigenous) growing carbon layer, the bog theory fails to explain also, because those seams extend horizontally in the same thickness and subsequently also on steep slopes upwardly. But normally, several coal seams exist as separate layers once upon the other. At the Ruhr (a river) in Germany, there are probably to account about 1000 separate coal stratum. which would correspond to the same number of forest bogs at the same location: One thousand continental raised bogs, each

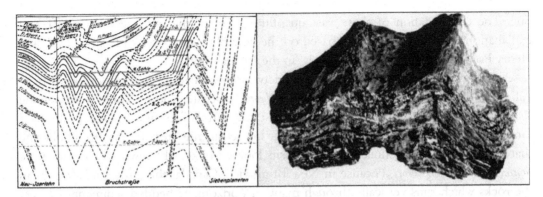

Figure 92: **Without cracks.** *Left picture: Folded seams are often extremely steep folded. The approximately 1000 m depth representing north-south cross-section of a mine field at Langendreer in Bochum, Germany, is showing a peaked folding directly located under flat-lying layers of fat coal (Kukuk, 1938, p. 322). Despite extreme folding the layers are mostly uniform (free of cracks). Therefore, there are two possibilities: On the one hand, these strata must have been plastically at the time of forming, or on the other hand, layer upon layer were directly deposited into this steep bedding. Both possibilities contradict the raised bog theory according as to Potonié, because subsequent folding of already hardened layers contradicts the strength of materials, because the hardened layers would tear and crush depending on the size of the superimposed load, tractive and / or shear forces. Right image: homogeneous wave folded coal layers of the coal mine Wehoven in Walsum, Germany.*

"With the regard to the absence of comparable phenomena in recent peat bogs, it is difficult to get a logical genetic-physiography interpretation of this irregularity of seams" (ibid, p. 252).

It should be examined, how electrochemical processes exert an influence, analogous to the description in this book.

up to 100 meters thick or more, and these once upon the other. An incredible idea! Complicating it comes in addition, that a not counterfeited carbon layer on one side sometimes separates into numerous coal seams, which are horizontally separated by layers of limestone and other formations (Figure 93, p. 225).

Individual coal seams, however, are sometimes 20 or more meters thick. No forest is able to produce such a massive carbon layer. It is estimated that the emergence of only 30 centimeters (12 inches) thick carbon layer, a four-meter-thick layer of peat would be required, and such peat layer would require a 40 meter high layer of plant remains. But then we must consider, how high and how dense a forest must be to produce a coal seam with thickness of 20 meters instead of 0,3 meters in thickness as calculated before? In some places, it must have been growing 50 to 100 of such unimaginable huge forests one above the other. And each of this forests would had been died back and buried one by one. The consideration of these vast quantities of plant materials requires a different explanation than those which are based on the very old hypotheses of Charles Lyell and Henry Potonié; brought forward already in the 19th century.

Also the deep black and lustrous variety of solidified hydrocarbons, known as Albertite, do not fit the *continental raised bog theory*. Albertite is less soluble in turpentine than the usual type of asphalt. That was first truly studied by the geologist Abraham Gesner, who had heard stories of rocks that burned in the area. It was from Albertite that kerosene was first refined. These deposits look coal-like, which would contradict the *continental raised bog theory*, because in New Brunswick, Canada, this carbon fills cracks in the rocks which runs vertically through many of horizontally bedded sedimentary layers (Hitchcock, 1865). Advocates of the organic origin described solid-hard Albertite as a type of asphalt. This should be the result of crude oil after the lighter hydrocarbons have evaporated because only a pasty or liquid mass would have to fill all cavities completely. However, as initially reported in this book, I had at my dinosaur research found that coal was formerly soft in the dinosaur era and formed bottom of lakes. Therefore dinosaurs left prints in sea grounds which build massive coal seems today. Many museums have exhibited such hardened coaly dinosaur prints.

The biogenic theory can't explain the fact that coal seams occur in places where such deposits geologically should not have to exist, and in angles of elevation that are not possible to allow a development of coal seams. But there are not only coal deposits deposited between layers of sediment, but also between volcanic lava without any sediment, including the southwest of Greenland (Pedersen / Lahm, 1970).

Also unusual are some lumps of carbonate rocks, which can be encountered in coal seams. If you break these chunks of rock, sometimes in its interior you can find fossils that contain light wood with no evidence of carbonization or process of coalification. According to the biogenic theory would expect black wood. It is documented in various coal seams not only some tree trunks, but several of those are standing upright and pierce multiple horizontal bedded layers – like in the Joggins Fossil Cliffs in Nova Scotia, Canada.

In the Donets Basin in Ukraine "can be found fossilized tree trunks that span through a coal seam from the carbonate rock below to that above. Those fossils are coalified where they are within the coal seam and are not coalified where they are in the carbonate" (Gold, 1999, p. 97). From this fact we can conclude that higher temperatures which are necessary for the coalification process analogues to the *continental raised bog theory* could not have been present.

Another, with in situ grown trees, not to explain fact is the occurrence of often vast area of coal deposits. Thus we find in the Appalachian coal field in eastern U.S. with coal seams, which extend over 132,000 square kilometers (km^2) with a constant distance to each other. Also the immense Pittsburgh coal field was estimated of about 50,000 km^2.

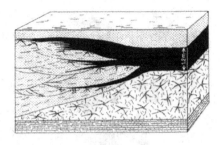

Figure 93: Divided. Coal seams which are not divided on the one side cuts into several small layers on the other side like a fork or fishtail shape, and then subsequently as to be replaced by a sandstone layer completely. Image: coal mine Wehofen, Germany (Kukuk, 1938, p 250).

Many researchers puzzled about silt and clay occurring of only a few centimeters in thickness, which pass through some coal deposits without interruption for several hundred kilometers horizontally and divide such horizontal seams (Wanlass, 1952). Up to date there is no satisfactory explanation of the origin of these sedimentary layers within coal seams in terms of the biogenic theory.

However, if such a layer of clay with very small pore spaces span above a stream of upward migrating hydrocarbons, then it acts as a filter. The heavy molecules of the hydrocarbons become stuck in the pores of the clay until the pore volume is completely filled. Finally, there is developed a bituminous mudstone, also called bituminous shale. This is not a slate in petrographic sense, but layered and not slated sedimentary rocks. Gladly is argued in the sense of the biogenic theory that this oil contained in the clay contains extinct plankton. Since the normal crude oil contains such biological remains this is deemed to be a conclusive proof for the organic or biological origin of petroleum and bituminous shale. Before we discuss this objection, finally, we turn to the peat and brown coal.

Except Peat and Lignite

Just as between black coal and crude oil, there is a close connection between peat and brown coal, but no intermediate stages of brown coal to black coal. For brown coal a basic role plays the organic material of extinct trees, also entire stumps, shrubs and grasses which was covered with various sediments and then may be geochemically converted under pressure and exclusion of air. Such a mixture is called peat, if the content of biological substances reaches 30 percent, otherwise it is called muck.

In peat and brown coal, the usual decomposition has been prevented so that the carbon was not converted into carbon dioxide and therefore do not get into the atmosphere. Thus a carbonaceous mud remains. The materials contained in that are stable over long periods of time and also reserve water at the same time. However, brown coal is in contrast to black coal geologically much younger and developed usually after the end of the dinosaur era in the geologic period Tertiary – the brown coal age.

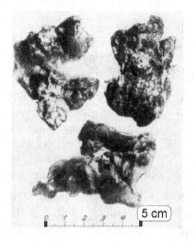

Figure 94: Ozokerite. This rare carbon-hydrogen mineral usually occurs as a thinly coating. Only a few cases of large deposits are known. In the coal mine Rheinpreußen I / II in Germany, it was found a large quantity of a few millimeters thick, milky-yellow ozokerite flakes, which fulfilled the fissures of the sand shale. The mineral was at 14.04 per cent of hydrogen and 85.03 percent of carbon, thus nearly pure carbon. "Any relation to oil deposits could not be found" (Kukuk, 1938, p 570). A very thin coating in fissures can hardly be almost pure carbon, which were caused by a process of coalification in the face of the raised bog theory according as to Potonié.

Because in these swamps is a lack of oxygen, microbes which feed on methane prevail against those which live on oxygen. Therefore it comes to the appearance of swamp gas that ignites itself occasionally. However, is the amount of biogenic methane formed in swamps or in brown coal fields high enough to make plausible an age of several million years? It has been shown frequently that out of brown coal fields escapes much more methane than biologically could be contained in this ever. This fact also applies to the black coal fields!

A characteristic time, for the biogenic formed methane to diffuse out of a two-meter thick coal seam into the surrounding rock is estimated to be about 10,000 years. Now if we apply to the coal deposit the 400-fold age respectively an age of 400 million years, then it is clear, that any methane may be not present in the coal seam after such a long time period. But never the less methane explosions are feared in many coal mines, becaus in spite of the very strong mine ventilations such disasters have occurred frequently. The coal mining on the Japanese island of Hokkaido even had to be discontinued because the best ventilated mines in the world were threatened by a latent danger of explosion.

Considering still in agreement with the conventional wisdom that the organic material must be heated in order to convert to coal, high temperatures had initially accelerated the production of methane very strongly. The diffusion would then be increased even early on a very high rate, and later on the amount of methane released in the coal deposit would be much lower.

How can be explained, the large quantities of methane escaping from coal seams these days, given the fact that there was formed not enough methane on biological way? Either these coal deposits are much younger than the geological time scale account, may be exhibiting an age at most of only a few thousand years, or there is always new methane

226

migrating in the coal fields. But exactly this scenario provides us with the theory of the abiogenic origin of hydrocarbons. In particular: methane migrates upward from the depths of the Earth and leak from fracture zones unnoticed in the atmosphere, if these hydrocarbons do not oxidized already before as it usually happens. The vertical stacking of horizontally overlying coal seams then confirms that took place a methane gas emission in this area over an extended time period respectively this process is still taking place, as long as there are conditions that favor the *deposition of carbon*.

In such a manner, a uniform or periodic gas leak over long geological time periods are maintained. On the contrary if it will be taking into account a biogenic carbon formation from plant remains, the amounts of carbon decline with time. In this case, the ratio of carbon to minerals is reaching such low levels as that at the surface vegetation never occur. If we think of tens of millions of years old coal deposits, then the high methane content, which is nowadays measurable, should almost entirely based on an inorganic (abiogenic) origin.

Particularly this become apparent in areas where is to note a not uncommon presence of peat and brown coal fields overlaying productive oil and gas fields, like in Sumatra and the Strait of Magellan. In this peat and brown coal fields just below the surface the gases were greatly enriched with methane, as it is commonly the case in such environments – a condition that supporters of the biogenic theory of course attribute to the presence of methane-excreting bacteria feeding on the plant debris in an oxygen-poor environment. But the gases were also enriched with all the other hydrocarbon gases from ethane (C_2H_6) to pentane (C_5H_{12}). "This mix is not normally produced by plants in any of their stages of decay. Quite simple. Microbes do not excrete pentane while decomposing carbohydrates" (Gold, 1999, p. 101).

Pentane is not a gas but a colorless liquid smelling like petrol. Interesting in this context is that occurrences of "white oil wells". There were "in the area of southern Iran in Achwas two wells, which delivered 75 liters of colorless oil each day" (Stutzer, 1931, p. 272). Also in Baku on the Caspian Sea Oil was quarried white crude oil. This field was small, and there was a limited flow rate. Because this oil has a very high quality, it was never exported. If one mix this white oil only with pure water, it can be used in some machines directly as fuel.

The diverse information, facts and theories for the abiogenic (inorganic) formation of hydrocarbons rising from the depths will be not discussed officially, because evidence seems incontrovertible: Petroleum frequently shows optical properties that arise as products of vital activity of organisms. Almost all of the naturally occurring crude oil contains unique decay products of organic molecules, which may impossibly have been formed in an inorganic process – is said categorically. If polarized light penetrates a sample of oil, the plane of polarization has turned to the left at the resurgence. The rotation is due to right-handed molecules, which are present in the oil in a different accumulation of the statistical mean. Therefore a direction dominate and is typical of biological fluids. This is not the case if the liquids developed abiogenic.

In oils not only a large number of different molecules are contained, of which some cause rotation to the right or contrary to the left, but in some other oils such a characteristic is missing completely. If oil were organic in origin, then they should feature a corresponding asymmetry analogous to the microorganisms, since most biological materials have a symmetry. However, some oils have not the smallest optical activity, which originate from different reservoirs at different temperatures of the same oil field, as demonstrated for the Lake Washington Field in Louisiana (Philippi, 1977). Crude oils with no optical activity testify either of a thorough destruction in terms of a biogenic origin of all chemical molecules *or of an abiogenic origin.*

How it is possible to explain alternatively the presence of dead microbes in crude oil? Actually, in the sense of the abiogenic theory the question appears to have been answered already: On the way up, so one might think, hydrocarbons sluice microbes, which are contained in sediment layers, in what way biogenic remains get in petroleum deposits from below.

However, some deposits of crude oil have no direct connection to a sediment layer, from which biological material could originate. But we have already established that microbes are living in incredibly large numbers in the pores and cavities of sediments and rocks, down to a depth of about six miles, because in these areas prevails a temperature at which such microorganisms are able to live. This from Thomas Gold (1999) forwarded idea seemed to be utopian a long time, because nobody could imagine that life exist at that depth. It was assumed that the rock would be pressed together seamlessly by the prevailing pressure down there. Therefore life was only possible at the surface, particularly in consideration of a hot Earth.

In 1984 appeared (to my knowledge) the first time an educated study of renowned chemists scientists from the University of Strasbourg, which confirmed that there is a rich biosphere with a large amount of biological material far down in the depths of the Earth (Ourisson et al., 1984). But these scientists assumed that these living bacteria itself had (organic) produced the oil and coal reservoirs. But then one must ask: from what nutrients (energy) are bacteria feed on, if not from hydrocarbon? However, for the production of so much oil would be necessary a very large number of micro-organisms, which in turn require massive amounts of (life) energy. Which source of energy should have been available in the depth, if chemical energy in the form of hydrocarbons discard?

Quite simple, we could come to the answer through the back door: no oil is produced by bacteria (except in individual cases), but oil is the food source for a rich bacterial life. This could disperse, because the hydrocarbons are streaming upward through the crust into the near-surface layers of the Earth and compose the source to supply life with chemical energy. Thanks to the photolysis, respectively splitting of water by sun light, plenty of oxygen was present, with contemporaneously volatilization of hydrogen up into space. Bacteria with the ability to oxidize methane (possibly hydrogen, carbon monoxide and water sulfide), may flourish in the crystalline rocks. This flora, which was obtained by outgassing alive, has left its particular biological materials respectively optical markers in petroleum and coal, wrote Thomas Gold (*Scientific American*, 251/5, p. 6).

Scientists search out, however, that biological materials in petroleum is a group of molecules that was called *Hapanoids* (ibid., 44–51). These were found in 2.7 billion year old Australian rock layers, making them one of the oldest evidence of life on this planet (Brock et al., 1999). There is no doubt that these molecules originate from the membranes of living cells.

Hapanoids in crude oil (petroleum) samples are represented very numerous, and Ourisson (et al., 1984) estimated the existence of at least 10^{13} tons. This is a 10-fold greater amount of organic carbon, as in all living organisms on the Earth's surface is contained with about 10^{12} tons. But to take into account is, that the researchers in 1984 are not differed between bacteria and archaea (*formerly called* archaebacteria). Interestingly, however, is a question that the researchers could not solve. On the one hand there was found that only bacteria form long-chain hydrocarbon molecules having up to 36 carbon atoms, but on the other hand such are not contained in trees, grasses and algae. The biological materials naturally occurring in hydrocarbons are altogether components of bacteria or archaea. However, they appear not in the flora and fauna from the Earth's surface. This means that life based on photosynthesis is not eligible to explain the presence of biological molecules in crude oil.

In principle, we knew this already in 1963 for other reasons, because it is highly unlikely that biological remains in hydrocarbons, which are saturated with hydrogen, can be converted. The famous chemist Sir Robert Robinson, president of the Royal Society in London was therefore considered that: Actually, cannot be underlined strongly enough that crude oil does not provide the picture of a composition which is expected from modified biogenic products, and any conclusions drawn from the components of old oils fit equally well or even better to the original idea of a hydrocarbon mixture, that was added organic products (Robinson, 1963).

No one has yet produced oil from marine organisms (plankton) or coal in a laboratory from a cup full of algae and fern!

The question remains as it is to clarify why crude oil is mostly containing remnants of dead microorganisms. This fact seems to confirm the biogenic theory. When Thomas Gold his theory about the *deep hot biosphere* published in 1992, this was declined. It was believed that in the hot depths could not live microbes. Those which were brought up to the surface with the drilling process, was declared as not to be resident down there, but as an impurity, which was rinsed with the drilling fluid from top to bottom down of the borehole (see Parkes / Maxwell, 1993). First, this argument was not contradicted.

Only then in September 1995, the argument of the contamination has been refuted scientifically, namely in the journal *Nature* (vol. 377, 21 September 1995, pp. 223.) It was shown that the microbes, found in a depth of 1600 meters in France, originate from a thermophilic community living down there (L'Haridon, 1995). Already in the following year was reported in the journal *Science* (vol. 273, 26 July 1996) that in the oil fields of Alaska biologically active respectively living microbes were found at a depth of 4200 meters at a prevailing temperature of 110 degrees Celsius (Fyfe, 1996). It goes even deeper: In Sweden it was discovered an active community in a depth of 5200 meters (Szewzyk, 1994).

Subsurface, where is to find crude oil, there are organisms flourishing. If one draw out crude oil together with living microbes, these die with the uplift process by no later than coming at the surface. The reason is simply that these organisms can only live under certain circumstances in the deep, where prevail special pressure and temperature conditions. One can compare this with deep-sea fishes which live in water depths of several kilometers. They also die as soon as this fishes gets to the water surface.

So, if we draw out petroleum from the deep, then this must contain dead microbes that nourish on petroleum as a chemical energy source down there, but they die off as a consequence of the decompression on the way up to the surface! The microorganisms are sometimes so abundant down there that the drill pipe is going to be clogged at the bottom – a horror scenario for drilling experts. When I visited the gas and oil fields northwest of Dallas, a drill pipe got blocked at the deep end of the borehole because the crude oil became dogged in case of too much microorganism which were living within the energy rich hydrocarbon layer (Photo 3). Therefore it is no evidence that crude oil has been generated biologically, although it is possible to find remains of dead microorganisms in crude oil.

To prove that crude oil contain *living* microorganisms and to study these, one must therefore remove a sample, seal it down there and cultivate the still living microorganism in a laboratory. With such samples taken in Sweden then also two previously unknown microbial strains have been discovered which proliferate only under the conditions prevailing below, in this case in the range of 60 to 70 degrees Celsius (Szewzyk, 1994). Of these microbes it was assumed rightly, that they nourish on hydrocarbons. However, it is officially assumed that these hydrocarbons are constitute transformed remains of organisms that once lived of photosynthesis at the Earth's surface. Accordingly, in the depths living microbes that nourish on hydrocarbons without oxygen, should be seen officially to be submerged with the biological material – in this case more than 5000 meters deep. This unbelievable scenario is owed to the supposedly biogenic origin of crude oil.

If one want to know more about the origin of life in the depths, than it is necessary to build corresponding containers that maintain the pressure and the temperature prevailing down there. Such experiments are very expensive and there are hardly existing experimental facilities that can simulate the survival of living microbes in up to ten kilometers deep. So far we have only a glimpse of the deep area of life. Probably we stand before spectacular discoveries, because chemical energy down there exist in various forms. Therefore there will be a corresponding variety of microorganisms that feed on different chemical energy and in turn produce various chemical products.

Because oxygen is the second most abundant element in Earth's crust, life exist down there as well as oxygen atoms that are weakly bonded to other elements. For example, if one takes away a part of oxygen from iron oxides, less oxidized iron with magnetic properties remains. These include also minerals such as magnetite (Fe_3O_4) and greigite (Fe_3S_4). If one takes away oxygen from the sulfates, then either remains pure sulfur or also sulfide, as for example iron sulfide. The presence of such metabolites in the deeper layers of the Earth's crust are indicative of biochemical processes that have taken place.

The living organisms in the deep must have been supplied by a uniformly flowing source of energy that microbes are not able to came at, because otherwise they would had plentiful proliferate and used up the entire supply of chemical energy rather quickly. This would have prevented their development. Therefore, the source of chemical energy (hydrocarbons) must to be located underneath the area of microbial life. As against the Earth's surface, in the depths are ruling relatively *constant* conditions, for which reason the evolution of microbes was able to evolve undisturbed over long geological periods.

The *deep hot biosphere*, which in the conventional geological concept represented a nonporous rock basement in a depth of some kilometers, had been clearly confirmed in recent years by discoveries of microbes living down there. Similarly, repudiated the idea that the source of chemical energy – in the form of hydrocarbons – could still be positioned below the range of microbial life. As opposed to this it was absolutely sure up to the millennium that hydrocarbons in very great depth of the upper mantle could not exist, because according to the theory of a hot Earth, the prevailing temperatures in this depths should be too high, for which reason hydrocarbons were decomposed. Therefore, the inorganic respectively abiogenic origin of hydrocarbons in the upper mantle had not considered seriously of most Western researchers.

But this opinion – represented scientifically since some decades – is changing last few years. Now it is believed, that at the upper mantle border, about 40 kilometers deep, shall dominate only temperatures of a few hundred degrees. But these temperatures are below the so-called *solidus temperature* up to this substances are fully present in a solid phase. These area is located in a depth of 60 to 210 kilometers below the Earth's surface according to updated geo-scientific conviction. Below this range with substances in the solid phase shall be located the asthenosphere, whereas the lower boundary extend at a depth of 300 to 410 kilometers. In this area the solidus temperature is exceeded. But down there, however, the materials should be partially melted only one to five percent.

Below the asthenosphere in fact no liquids shall exist, but more harsh and brittle material may be present. So, my idea is confirmed, because I'd visualized the view of an only partially molten mantle in my book *Mistake Earth Science* (Zillmer, 2007) – original published in Germany 2001. So it is on hand the requirement for hydrocarbons to migrate upward from deep sources! Because the mostly material in the asthenosphere is not melted, it is nowadays accepted that there is contained a certain amount of water in the rocks. A few years ago this exposition was considered to be fancifulness.

An important postulate for the accuracy of the abiogenic theory was confirmed. So hydrocarbons are streaming up from great depths? According to recent research findings that question will be positively answered by some professionals. Taking into account the scientific discussion of the last 40 years, then we are surprised, even amazed!

So recently, the existence of stable carbon is suspected in depths of about *400 kilometers* – confirmed in a dissertation from 2006 in Germany (Shcheka, 2006). This carbon-excess in the Earth was "from chemical data of the cosmos derived, based on the observation in meteorites and other objects in the solar system that in spite of the relative reduction of volatile substances, the ratio of these to similar others substances behaves often independent proportional to the degradation process" (ibid, p. 6).

This seemingly curiosity and the apparent relative constancy of the disproportionately large supply of volatile substances in comets was confirmed by recent satellite researches. This fact was also based on my research over the years . We have interpreted this as confirmation of a carbon source in the interior of comets, planets, moons, and even of the Earth, especially because newly is accepted the existence of carbon in the depths of the upper mantle and therefore in the *deep hot biosphere*.

Of course, the officially propagated carbon cycle can hardly take place through the granitic continental plates. Therefore, it is newly thought, that carbon is sinking into the mantle and this with the oceanic tectonic plates which allegedly dip into the subduction zones. As already discussed in my book *Mistake Earth Science* in detail, there cannot exist subduction processes (Fig. 6, p.20) because on the one hand, the rock materials of the subduction plate are lighter than the material in which this should be deepen, and on the other hand, because the subduction plate would tear before these could sink in case of statically and material-technically reasons (see Fig. 7, p. 22).

The alleged regeneration of carbon through subduction into the mantle was already presented in the scientific journal *Nature* in 1982 (vol. 300, 11 Nov 1982, pp. 171). It is justified: This conclusion is reached by comparing of integrated materials upward-streaming from the mantle and the presence of independent carbon reservoirs (Javoy et al, 1982). After all, this reasoning corresponds to that in this book developed view of upward-streaming (hydro)carbon. But because in these great depths cannot be an independent source of carbon, as most researchers be convinced, because they think that carbon must be countersink 400 km and more, so that the *observed carbon balance* can be

Figure 95:
Thermochemical
calculations.

Figure A shows the concentrations at 500 degrees Celsius. At pressures of more than 0.9 gigapascals (GPa) originate methane.

Figure B shows that at temperatures of more than 1500 degrees Celsius molecular hydrogen (H2) and carbon dioxide is dominant instead of methane.

Liquid state = (F)
Solid state = (S)
According to Scott et al., 2004.

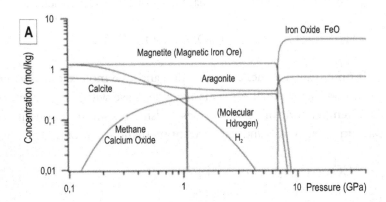

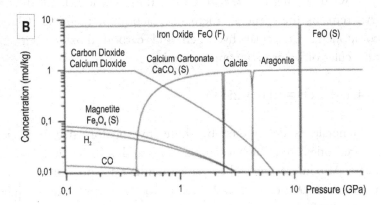

maintained (Shcheka, 2006, p. 9). I underline the word balance, because most researchers believe in a constant amount of carbon and thus a self-contained cycle.

Some scientists have disagreed with this opinion (including Marty / Jambon, 1987 and Cartigny et al., 1998). Then appeared in September 2004, the description of intensive laboratory tests, which were attended by scientists from several reputable institutes and laboratories in the United States (Scott et al, 2004): This study demonstrates the possibilities of the abiogenic formation of hydrocarbons at pressures prevailing of five and eleven gigapascal and temperatures of 500 to 1500 degrees Celsius in the upper mantle of the Earth. It is presumed valid, that the carbon content of the soil could be larger than conventionally assumed (ibid, p. 14023).

The calculations of this study predict that the concentrations of calcium oxide (CaO), respectively burnt lime, and methane (CH_4) are almost identical, prevailing at a temperature of 500 degrees Celsius. From this it can be concluded of the decomposition of calcium carbonate ($CaCO_3$), if (molecular) hydrogen (H_2) is added (ibid, p. 14025), that is upward-streaming from the mantle according to the discussion in this book.

$$4 H_2 + CaCO_3 = CH_4 + CaO + 2H_2O \quad (A)$$

It originate methane, calcium oxide and two molecules of water. The calculations also show that the production of methane decrease with the gradient of higher temperatures. At a temperature of 1500 degrees Celsius results in a roughly equal concentration of calcium oxide (CaO) and – instead of methane – carbon dioxide (CO_2):

$$CaCO_3 = CaO + CO_2 \quad (B)$$

If we now subtract the high-temperature reaction (B) of the low-temperature reaction (A), remains the (general) methane reaction of methane (CH_4), which consist of the components molecular hydrogen and carbon dioxide; but also are originated two molecules of water in the upper mantle:

$$4H_2 + CO_2 = CH_4 + 2H_2O$$

Generally is determined by these investigators that hydrogen and carbon dioxide favored originate at higher temperatures and at lower temperatures mainly methane and water. Also interesting is the observation that it is possible to form methane bubbles with a reduction of pressure and cooling to room temperature. The researchers conclude (Scott, 2004):

The analysis shows that the production of methane benefits in a wide range of high-temperature conditions. In this study has documented the stability of methane over a wide extending pressure-temperature range. The outcome of this is the *plentifulness of hydrocarbon* of our planet and furthermore it is pointed out that methane is much more prevalent for the storage of carbon in the mantle than previously thought – wherewith details are linked to the existence of a *hot deep biosphere*. Although isotopic evidence shows the wide distribution of hydrocarbon gas reservoirs in large sedimentary basins, but these observations and analyzes do not include the large abiogenic reservoirs in the mantle.

Rather, the assumption that carbon dioxide is the solitary carrier of abundant gases escaping from the mantle will be reviewed and re-evaluate. Finally, the potential for the originating of heavy hydrocarbons under high pressure conditions may exist, and whereat in the mantle originated methane is to see as a precursor (ibid., pp. 14026).

The originating of hydrocarbons in the mantle and its upward-streaming from great depths is considered to be realistic due to these thermodynamic calculations. In the area of the sediment package, overlying the crystalline basement rock, these sediment layers are systematically enriched with carbon, whereat on the one hand hydrogen is released on its way up and on the other hand will be originated water and carbon dioxide as a result of an oxidization process. So we get a picture of a systematically layering which overlies the crystalline basement rock.

Also the fact that coal deposits are positioned above petroleum deposits, is hardly to explain by the biogenic theory. Also, an answer cannot be given on the question of why coal strata once formed lake grounds and therefore dinosaurs could had sank into these coal layers, which were originated relatively short time before and therefore the stone coal was still soft in those places! This irrefutable fact is contrary to the conventional theory in regard to a nearly endlessly continuing coalification process which require high temperatures and pressures. But there any overlying top layers of rock or soil were missing and *therefore no high pressure could built up*, only a very less pressure due to some meters of water coverage.

There is another little discussed aspect, because coal contains substances whose presence cannot be scientifically explained if it is to consider a biological origin and a coalification process. So, black coal and petroleum products thus contain up to four percent sulfur. Therefore, sulphur dioxide originates during the burn-up "fossil" fuels. But there is also a unconventional unexplained part of the coal.

The Radiation of Black Coal

If we pose the question, where more radiation prevails, in the vicinity of a nuclear or yet a coal-fired power station, in all likelihood is remarked a wrong answer. This radiation dose – additional to the natural radioactivity – which get hurt the population in the vicinity of a coal power plant is significantly higher than for a nuclear power station, partly because coal contains quite a lot of uranium, even up to 60 grams per ton of coal. The radioactive noble gas radon as a decay product of uranium passed without hindrance all fume gas filters, and thus enters the atmosphere.

In the coal ash produced by the burn-up of coal, the concentration of uranium is then more than ten times higher than in the original coal, because the uranium content is increased during the burn-up process. In this way we obtain concentrations as they are typical for some uranium-rich ores that are mined. Uranium containing ash dumps in some places may therefore represent a threat to groundwater. It would make sense, the coal ashes directly to use for the extraction of uranium. On the one hand, the ash dumps were refurbished and on the other hand could then be used in uranium nuclear power stations. The half-life period of (radioactive) nuclides contained in the ash would be greatly reduced in the reactor, and there remains only a small reside that must be stored in a nuclear waste repository.

In addition to uranium there is thorium contained in coal and this in an even greater extent. Thorium is used in nuclear power stations as a primary energy source. According to estimates by the *Oak Ridge National Laboratory* in Tennessee, with the use of coal would be laid off 828,632 tonnes of uranium, therefrom 5883 tons of uranium-235, and 2,039,709 tons of thorium during a time period of one hundred years from 1940 to 2040 (Gabbard, 1993).

A Canadian mining company wanted to get about 120 tons of natural uranium each year from the constantly accumulating ashes of three coal-fired power stations in the province of Yunnan in China. Even in Europe there are existent enough such uranium sources. For example, can be found in the Czech Republic and Hungary brown coal (lignite) with comparable levels of uranium.

Today, because there are plenty amounts of uranium – also in seawater – these days and also in the future, it is *not* necessary to open up new uranium deposits. For example, if at the power plant inlet structure of the tidal power station in La Rance od Northwest France one would be extracted only 20 percent of the actually in seawater dissolved uranium, then you could run a nuclear power station of 1500 megawatts of electricity, that produces 25 times m more than the tidal power station itself (Prasser, 2008).

Now then, in black coal there is included a partially very high uranium content. If you believe that black coal was actually resulting from organic material, raises the question of how uranium has got into lignite and black coal, or alternatively already into the allegedly pre-existing peat bogs? There have been made studies on coal strata in the Colorado Plateau, where dinosaurs were running around on soft Cretaceous coal layers. Thereby was interesting to note that decay processes of a single atoms took place in different locations. In several coal deposits the researchers found a lot of *isolated* polonium 210-nuclides, but which are the at last formed stages of radioactive nuclides in the decay series of uranium-238, before the finished product originates in the form of stable lead.

Each stage of the radioactive decay of an uranium atom can be detect in a hardened matrix such as coal, since *each* (radioactive) nuclide leave a halo with a characteristic diameter in the form of a three-dimensional spherical respectively ball-like radiation effect. When polonium is found without its radioactive decay precursors in the form of their special halo, than this nuclide must have been transported away from the original site of uranium decay into a different substrate, may be in this case into a coal deposit. This must have taken place before the coal layer hardened, because polonium-210 decays quickly and only half of the original amount is present in just 139 days. Long distances cannot be covered by a slow diffusion process. Therefore, a movement of polonium must have happened within a time period, when the substrate was to be arranged mostly in a squashy or at maximum plastic-deformed state. Therefore the hardening process must be finished in a period that lasted only months but not years and the infiltration of polonium must be occurred quickly, as the "coal" was still in a soft state – as a study shows published in 1976 (*Science*, vol. 194, 15 Oct 1976, pp. 315).

Figure 96: Chattanooga Shale. This native rock, encountered over large areas of North America, has a film-like appearance and is heavily mixed with carbon. Until now, there take place no commercial exploitation of the Chattanooga Shale, although it is regarded as a source of oil and gas fields in Tennessee, and is suitable for the production of hydrocarbons.

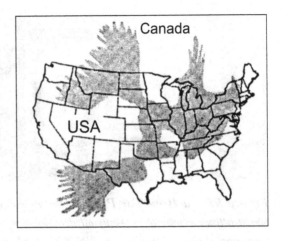

It follows that this coal hardened quickly and not just "imperceptibly slowly", according to Lyell's geological hypothesis. In addition, it is also unlikely that plant material could have formed a plastically-deformable mass as a pre–condition to start the coalification process.

The inorganic and just *not organic* origin of black coal would also assisted by the question, whether hydrocarbon fluids serve as an upward carrier of uranium atoms. Actually in this way it is possible, that polonium 210-nuclides can be washed away from their parent nuclides with a quick upward cascade-like pumping action of hydrocarbon fluids. Finally some hydrocarbon bulks hardened at the surface and not underneath a cap rock. Therefore dinosaurs and other animals could leave tracks at top of the developing black coal deposit; but also below this carbonic layer before the coal layer started to originate. But *inside* the softly coal seam were located polonium nuclides, which were unable to crack open rock crevices by itself, but these gushed out with the fluids from bottom up.

The "young" age of these coal strata is confirmed, as was discovered far too little lead in relation to uranium as the source, because lead it is the final and stable stage of decay chain. To check whether this is an exception, alternatively, the carbonaceous Chattanooga Shale (shale) formation was investigated, which extends over large areas of North America. In this layer, there was determined a ratio of uranium-238 to lead-206 resulted with a factor of 1000 too high to correspond with the alleged high geological age of origination (Gentry et al., 1976). But according to the geological timescale this Chattanooga shale should be about 360 million years old (Devonian). Considering that only 2.5 percent of the uranium atoms in general have an halo appearance *of any* decay stage, the researchers concluded that possibly both, the infiltration of uranium and the subsequent coalification within the last few thousand years has taken place (ibid, p. 317).

Figure 97: La Brea Tar Pits. At the beginning of the 19th Century in the area of present-day tourist attraction were to see many oil derricks. The right image shows a block with a jumble of fossil bones from the asphalt exploitation "Pit 8" at La Brea as of in the year 1914.

Because dinosaurs were running around on young softly coal layers, Indians have certainly seen these animals in swamps and lakes. There are actually prehistoric drawings of dinosaurs. The traditions of the Navajos report that their ancestors and dinosaurs lived together at the beginning of the world.

Death Trap Asphalt Pits

In addition to the observed squashy black coal at the bottom of once shallow waters can be observed swelled hydrocarbons at the surface in the form of petroleum seep deposits. So there were holes that were filled with asphalt as to see in the oil region in Wietze near Hanover in Germany. This was still carried to end about 1900. Better known are the tar pits with natural asphalt at La Brea in Southern California, where still many oil rigs were visible in the Salt Creek Oilfields formerly outside of Los Angeles at the beginning of the 20th Century. Already before, thousands of tons of natural asphalt were produced and transported to San Francisco. Layers of oil shale in many places have a thickness of approximately 700 meters. Where shopping centers are located today there previously have sunk about 60 species of mammals, more than 2,500 saber-toothed cats, among others big cats, mammoths, mastodons, wolf-like giant dogs, American lions (Panthera leo atrox), jaguars, cougars, bobcats, coyotes, pygmy antelope, turkeys, elephants, big giant sloths and the extinct American camels and horses (Equus occidentalis), and birds, including peacocks, but oddly enough fishes, amphibians, reptiles, molluscs, insects, spiders, many plants and their pollen and seeds. These land animals were bogged down, because they had mistaken asphalt ponds with water lakes or only, because they got accidentally into this asphalt swamps? And what is with the aquatic life which bogged down also in the asphalt pits?

In fact, the animal remains in the asphalt pits are crowded and mixed together in unbelievable piles. During the first excavation it were discovered an average of 20 sabre–toothed cats and wolf skulls per cubic meter (Merriam, 1911). To date, more than 166,000 bones were discovered. As the majority of the animals should have been bogged down in the tar pits, before my visit to La Brea I was expecting that more or less complete skeletons were found in the asphalt; but this is only an exceptional case (ibid, p. 212). Actually the bones are well preserved, but they are fragmented and mixed to form a highly heterogeneous material. These cannot be taken place as the result of an accidental bogging down of around stroking animals.

Is it possible that in this area hydrocarbons shot up in the sky like a cascade? There, whereabouts herds of frightened animals, especially carnivores, were overwhelmed by a natural disaster in La Brea, and the landscape was ravaged by forcible storms and stones were turned into bullets? Sprang up hydrocarbons like a fountain hundreds of feet into the air after the caprock broke suddenly due to high pressure? Were the animals overwhelmed with falling bitumen and bogged down in it?

There were also found 17 bones of a woman which has been dated to an age of about 9000 years. The skull shows no differences from those of today's Indians.

The asphalt came suddenly to the surface and partly hardened very quickly, so that animals could ran on such layers at the surface, like the described dinosaurs were running around on the partly soft coal layers. Such a scenario, one can now observe at the largest asphalt lake in the world, at a place called La Brea too, but this is located on the Caribbean island of Tobago. From a crater, whose diameter is 1,500 meters, more viscous asphalt, natural swells up. It hardened rapidly at the surface, so it is possible to run on many parts of the lake. Only a quarter of the asphalt surface is still liquid. Along the edges it is possible to subside a bit, like the dinosaurs in the coal.

On the bubbling mass of liquid asphalt, natural methane escapes and produces dazzling colorful bubbles, and in some spots of dry areas seeps out oil which overflow over the asphalt surface as a thin film. The actual depth of the asphalt lake is still unknown. Here it is to see how the Earth is really active. Not only in the crater area, but also in the town center of the nearby village of La Brea asphalt gushes out of the Earth surface.

Out of the Dead End

The documented findings of dinosaurs that walked on soft coal, and the massive presence of hydrocarbons and volatile organic compounds in many celestial bodies in our solar system led to a radical re-evaluation not only of geological but also of astronomical scenarios. Modern plasma physics in the future will replace the old dogmas of astronomy, because it was not possible to explain the modes of action in the universe by purely mechanistic principles. Quite the contrary, by the doctrine of gravitation more riddles were created than solved.

The universe regarded to be functional exclusively in a mechanical mode of action, already Albert Einstein had difficulties to explain the cosmos. With his special theory of relativity on the one hand, he demonstrates an ether in the universe, to this then to deny. Originally, in his famous formula Einstein formulated not power, but instead of this electricity is equal to mass times the speed of light squared. Taking into account that gas in the universe is to be arranged in the plasma state, then electricity or energy can be transmitted through the (almost) empty space. So all electrically conductive celestial bodies are interconnected.

In this way, our Earth receives energy from the Sun and is controlled by it. Understanding the mechanics and the flow of energy in the depths of the Earth is helpful for us to find the reasons for the peculiar distribution of oil and gas on Earth. Understanding the processes inside the Earth and a view of the world as realistic as possible are essential to identify the current radical-extreme climate policy as such. Only then we are able to avoid the depletion of our society just in case of explosion in costs for energy not only in Germany (heating. fuels, electricity) which is based on false climate theories and artificial shortage of resources respectively so called "fossil" fuels. But nevertheless, it is to agree with a reasonable protection of our resources, expedient energy savings and meaningful environmental protection – if only for the reason to make us becoming independent of the resource-producing countries like the Middle East or the today independent states of the former Soviet Union. But this has nothing to do with climate change. Unfortunately, environment protection themes are justified always with climate protection and human-induced climate change, although an additional contamination of the atmosphere will lead to cooling down and not just warming. Climate change should be rejected, because the Sun controls our climate, and it should be become warmer rather than colder in the long run, although in the near future a slight increase in temperature seems possible to the point of the temperature level as it prevailed during the Medieval Climate Optimum. This is the time period as Vikings on Greenland cultivated grain and held more than one hundred dairy cows on their farms.

It is claimed that the world is facing a climate catastrophe due to a global warming, although it was several times as warm as today in the past 1,000 years. As the justification of the alleged human-induced climate change has to serve the increase of carbon dioxide in the atmosphere, which, according to climate activists, shall persisted for more than 1000 years at a consistently low level. But, the climate curve presented by *the Intergovernmental Panel on Climate Change* (IPCC) of the United Nations is wrong,

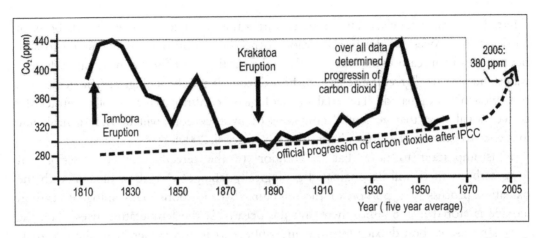

Figure 98: Carbon Dioxide progression. *The alleged by climate scientists, supposedly clearly associated trend (dashed line) coming along with the grade of industrialization does not exist. According to Beck, 2007.*

As the chemist Professor Hans-Eberhard Heyke already pointed out some years ago. The biologist Ernst Beck then created an extensive study. Here, 90,000 precise values were considered (Beck, 2007) – compare Figure 98.

The IPCC used at all, only about ten percent of the 390 since 1800 existing scientific works, of which were acknowledged only three to be correct, since only theses delivered exactly the values with which could be "proved" the greenhouse theory. In addition, it was found that two of the three works are considered wrong. The IPCC carbon dioxide curve should be the evidence that the greenhouse effect is anthropogenic respectively caused by humans, but is based on sloppy research and data selection. Ernst Beck proceeded with great care, since the IPCC already exerts a great influence on politics. Taking into account the total carbon dioxide of about 90,000 measurements showed that the carbon dioxide follows the climate and not vice versa, as claimed by the IPCC. Many results from ice cores are simply misinterpreted, as established in detail in my book "Mistake Earth Science," and a human-induced greenhouse effect is not detectable. This new study has uncovered the climate scientists affair, because all citations can be checked.

The degassing of the Earth is an unrecognized major factor in our conventional view of the world. The hydrocarbons ethane and propane are rising undetected in the atmosphere. According to a study published in January 2009, the up to date overlooked proportion of ethane and propane will cover five million tons, representing 50 percent of the previously considered amount which plants, bacteria and humans should emit, referred to climate activists. This study confirms the principle discussed in this book of an degassing process of the Earth, although the proportion of methane is still to add, because this hydrocarbon oxidizes already in the Earth's crust for the most part, whereby water and also the "greenhouse gas" carbon dioxide is purely geological (abiogenic) and not biogenic originated. The Italian geologists confirm: Our results refute the conventional view that geological emissions of hydrocarbons will be only a negligible influence to exert on the atmosphere (Etiope / Ciccioli, 2009).

It is important to note that in addition to the directly emitted emissions of hydrocarbons there will be originated geologically (abiogenic) carbon dioxide with the oxidation process of a portion of the migrating hydrocarbons. This abiogenic carbon dioxide is stored in the atmosphere and the oceans. If the temperature rises, then the oceans release carbon dioxide into the atmosphere. If it gets colder, then the oceans is going to store "greenhouse gases".

Also, methane itself, which is allegedly caused to 70 percent of mankind according to climate activists, is always emitted in the atmosphere directly with methane eruptions and through many fractures in the crust, some only with regional, but partly also with global consequences. This means that large islands can come into being or sinking fast, like the isle near Iceland as already described. It can also be transformed whole areas completely – sometimes accompanied by mass extinction of animals and humans. Many small and large scale disasters, such as the Christmas 2004 tsunami, can originate from gas eruptions. Even some previously misdiagnosed specific landforms can be traced back to gas rise. It seems important at all gas emissions to consider and to detect as early as possible, even if only to prevent plane crashes or tsunamis.

But more important that the human have effect on our climate is very little and barely measurable. The reality is that *the Sun controls our climate*! The currently fueled fears of a climate catastrophe is used to make large parts of the population amenable to higher taxes and charges and this mostly without contradiction. From the political perspective only threats and a fear propaganda will be best placed to attain one's ends, whereas it is a fact that only our offspring live to witness the predicted climate warming. But at this time the today's false predictions are long forgotten and those responsible should be retired. The climate policy is not science but about politics and making money (allowances), as the political demagogue Al Gore shows us so well.

Unfortunately, such researches that proves the warming of the Earth bred by humans, has become a lucrative source of research funds, which is why more and more researchers change sides and establish scientifically researches with climate topics which, in fact, are mostly environment topics. Each of the approximately 2,000 American climate scientist should be received an average of one million dollars per year for his research. Worldwide, the annual budget for climate research reaches about five billion dollars. The "big bluff" is accelerated in time, because you will notice no later than in a few years that the predictions about climate change should not take place. Already, down to the present day the average temperature of the air has not become warmer since 2000!

The oil reserves at that time will be only slightly lower than today, and natural gas resources will continue to increase. The supply of humans with sufficient energy is also guaranteed in the future, if the energy supply is not calculated reduced and in this way too expensive. To avoid a blackmail and to prevent economic and political dependence in relation to other countries, it is required to accelerate new technologies of energy production and wherever possible regenerative energies to make these techniques economically viable despite of still available great quantities of "fossil" energies for a long time.

For the above reasons and ultimately for a secure future for mankind, it is important to recognize the energy-mistake as such ...

Bibliography

Aculus, R. J., und Delano, J. W.: »Implications for the primitive atmosphere of the oxidation state of Earth's upper mantle«, in: »Nature«, vol. 288, 6. 11. 1980, p. 72 ff.

A'Hearn, M. F., et al.: »Deep Impact: Excavating Comet Tempel 1«, in: »Science«, vol. 310, 14. 10. 2005, p. 258–264

Anders, E., et al.: »Organic Compounds in Meteorites«, in: »Science«, vol. 182, 23. 11. 1973, p. 781–790

Anderson, D. L., und Dziewonski, A. M.: »Seismische Tomographie: 3-D-Bilder des Erdmantels«, in: »Die Dynamik der Erde«, Heft: »Spektrum der Wissenschaft: Verständliche Forschung«, Heidelberg 1987, 2. Ed. 1988

Anthony, H. E.: »Nature's Deep Freeze«, in: »Natural History«, September 1949, p. 300

Archer, D.: 2005. »Destabilization of Methane Hydrates: A Risk Analysis. Externe Expertise für das WBGU«, in: Sondergutachten »Die Zukunft der Meere – zu warm, zu hoch, zu sauer«, unter: http://www.wbgu.de

Babaev, E., et al.: »A superconductor to superfluid phase transition in liquid metallic hydrogen«, in: »Nature«, vol. 431, 7. 10. 2004, p. 666 ff.

Backus, G.: »Dynamo model at a turning point«, in: »Nature«, vol. 337, 21. 9. 1995, p. 189–199

Balsley, J. K., und Parker, L. R.: »Cretaceous Wave dominated Delta, Barrier Island, and Submarine Fan Depositional Systems: Book Cliffs east central Utah«, in: »Amer. Assoc. Petrol. Geol. Field Guide«, 1983

Barbos, J. A., et al.: »New dyrosaurid crocodylomorph and evidences for faunal turnover at the K–P transition in Brazil«, in: »Proceedings of the Royal Society B«, DOI 10.1098/rspb.2008.0110, 25. 3. 2008

Barker, C., und Dickey, P. A.: »Hydrocarbon habitat in main producing areas, Saudi Arabia«, in: »Americ. Assoc. Petrol. Geol. Bull.«, vol. 68, 1984, p. 108 f.

Bastian, H.: »Du und die Vorzeit«, Frankfurt 1959

Baumgardner, J., et al.: »Imaging the sources and full extent of the sodium tail of the planet Mercury«, in: »Geophysical Research Letters«, vol. 35, 2. 2. 2008, L03201

Baumjohann, W.: »Die Erdmagnetosphäre«, in: Glassmeier/Scholer, 1991, p. 105–118

Beck, E.: »80 Jahre CO2-Gasanalysen der Luft mit chemischen Methoden«, in: »Energy & Environment«, vol. 18, 2/2007

Belloche, A., et al.: »Detection of amino acetonitrile in Sgr B2(N)«, in: »Astronomy Astrophysics«, vol. 482/1, 2008, p. 179–196

Berckhemer, H.: »Die Grundlagen der Geophysik«, Darmstadt 1990, 2. Ed.1997

Berg, H. C.: »E. coli in motion«, New York 2003

Berner, U., und Steif, H.: »Klimafakten. Der Rückblick – Ein Schlüssel für die Zukunft«, Hannover 2004, 4. Ed.

Bilham, R.: »A Flying Start, Then a Slow Slip«, in: »Science«, vol. 308, 20. 5. 2005, p. 1126 f.

Bindemann, I. N.: »Die Urgewalt der Supervulkane«, in: »Spektrum der Wissenschaft«, August 2006, p. 38–45

Bostick, W. H.: »Experimental Study of Plasmoids«, in: »Physical Review«, 1957, vol. 106, p. 404

Brasier, M. D., et al.: »Questioning the evidence for Earth's oldest fossils«, in: »Nature«, vol. 416, 7. 3. 2002, p. 76–81

Brocks J., et al.: »Archean molecular fossils and the early rise of eukaryotes«, in: »Science«, vol. 285, 13. 8. 1999, p. 1033-1036

Buffet, B., und Archer, D.: »Global inventory of methane clathrate: sensitivity to changes in the deep ocean«, in: »Earth and Planetary Science Letters«, vol. 227, 2004, p. 185–199

Burckhardt, B., et al.: »Ikaite tufa towers in Ikka Fjord, southwest Greenland: their formation by mixing of seawater and alkaline spring water«, in: »Journal of sedimentary research«, vol. 71, 2001, p. 176–189

Cartigny, P., et al. (1998): »Eclogitic diamond formation at Jwaneng: No Room for a Recycled Component«, in: »Science«, vol. 280, 29. 5. 1998, p. 1421–1424

Chekaliuk, E. B.: »The thermal stability of hydrocarbon systems in geothermodynamic systems«, in: Kropotkin, 1976, p. 267–272

Chorlton, W.: »Eiszeiten«, Gütersloh 1983

Cifci, G., et al.: »Deep and shallow structures of large pockmarks in the Turkish shelf, eastern Black Sea«, in: »Continental Shelf Research«, vol. 23, 2003, p. 311–322

Collett, T. p.: »Permafrost-associated gas hydrate accumulations«, in: Sloan, E. D., et al.: »International Conference Natural Gas Hydrates: Annals of the New York Academy of Sciences«, vol. 715, 1994, p. 247–269

Cremonese, G., et al.: »Neutral Sodium from Comet Hale-Bopp: A Third Type of Tail«, in: »The Astrophysical Journal Letters«, vol. 490, 1997, p. L199–L202

Csányi, V.: »Evolutionary Systems and Society: A General Theory of Life, Mind, and Culture«, Durham 1989

Dando, P. R., et al.: »Gas venting from submarine hydrothermal areas around the island of Milos, Helenic Volcanic Arc«, in: »Continental Shelf Research«, vol. 15, 1995, p. 913–929

Dansgaard, W., et al.: »One Thousand Centuries of Climatic Record from Camp Century on the Greenland Ice Sheet«, in: »Science«, vol. 166, 17. 10. 1969, p. 377–381

Darwin, C.: »The Origin of Species«, London 1859; deutsch: »Die Abstammung der Arten«, Lizenzausgabe Köln 2000

Dawson, F. R.: »Hydrogen production«, in: »Chemical & Engineering News«, vol. 59 (15), 1981, p. 2

Decker, R., und Decker, B.: »Die Urgewalt der Vulkane«, Weyarn 1997

Deinzer, W.: »Die Sonne«, in: Glassmeier/Scholer, 1991, p. 1–16

Dillow, J. C.: »The Water Above: Earth's Pre Flood Vapor Canopy«, Chicago 1981, p. 371–377

Drujanow, W. A.: »Rätselhafte Biographie der Erde«, Moskau 1981 (russ.), Leipzig 1984 (German)

Egorov, N. N., et al.: »Present-day views on the environmental forming and ecological role of the Black Sea methane gas seeps«, in: »Marine Ecological Journal«, vol. 2, 2003, p. 5–26

Escher, B. G.: »L'éboulement pré-historique de Tasikmalaja et le volcan Galounggoung (Java)«, Reprint Leidsche Geologische Mededeelingen, 1, 1925, p. 8–21

Etiope, G., und Ciccioli, P.: »Earth's Degassing: A Missing Ethane and Propane Source«, in: »Science«, vol. 323, 23. 1. 2009, p. 478

Evans, p. E., et al.: »Biological Evans-Evolution«, in: »Proceedings of the National Academy of Sciences«, vol. 105, 19. 2. 2008, p. 2951–2956

Ewert, K. D.: »Die physikalischen Zwangläufigkeiten des Kosmos«, Haselünne 1985

Fahr, H. J.: »Der Urknall kommt zu Fall«, Stuttgart 1992

Fenton, L. K., et al.: »Global warming and climate forcing by recent albedo changes on Mars«, in: »Nature«, vol. 446, 5. 4. 2007, p. 646–649

Fischer, H.: »Rätsel der Tiefe«, Leipzig 1923

Fleischmann, M., et al.: »Electrochemically Induced Nuclear Fusion of Deuterium«, in: »Journal of Electroanalytical Chemistry«, vol. 261, 10. 4. 1989, p. 301

Fortey, R.: »The Cambrian Explosion Exploded?«, in: »Science«, vol. 293, 20. 7. 2001, p. 438 f.

Friis-Christensen, E., und Lassen, K.: »Length of the Solar Cycle: an Indicator of Solar Activity Closely Associated With Climate«, in: »Science«, vol. 254, 1. 11. 1991, p. 698 ff.

Fritsch, W., und Meschede, M.: »Plattentektonik. Kontinentalverschiebung und Gebirgsbildung«, Darmstadt 2005

Fuller, M. L.: »The New Madrid earthquake«, in: »U. p. Geol. Surv. Bulletin«, 494, 1912

Fyfe, W. p.: »The Biosphere is Going Deep«, in: »Science«, vol. 273, 26. 7. 1996, p. 448

Gabbard, A.: »Coal Combustion: Nuclear Resource or Danger«, in: »Oak Ridge National Laboratory Review, vol. 26, Ausg. 3/4 1993

Galimow, E. M.: »Isotopic composition of carbon in gases of the crust«, in: »Internat. Geol. Rev.«, vol. 11, 1969, p. 1092–1103

Gams, H., und Nordhagen, R.: »Postglaziale Klimaänderungen und Erdkrustenbewegungen in Mitteleuropa«, München 1923

Garcia-Ruiz, J. M., et al.: »Self-Assembled Silica-Carbonate Structures and Detection of Ancient Microfossils«, in: »Science«, vol. 302, 2003, p. 1194

Gary, p. P., et al.: »Computer simulations of two-ion pickup instabilities in a cementary environment«, in: »J. Geophys. Res.«, vol. 93, 1988, p. 9584–9596

Gentry, R. V., et al.: »Radiohalos in Coalified Wood: New Evidence Relating to the Time of Uranium Introduction and Coalification«, in: »Science«, vol. 194, 15 10. 1976

Gilette, D. D., und Lockley, M. G.: »Dinosaur Tacks and Traces«, Cambridge University Press 1989

Glasby, G. P.: »Abiogenic Origin of Hydrocarbons: An Historical Overview«, in: »Resource Geology«, vol. 56, 2006, p. 85–98

Glaser, R.: »Klimageschichte Mitteleuropas«, Darmstadt 2001

Glassmeier, K.-H., und Scholer, M. (Hrsg.): »Plasmaphysik im Sonnensystem, Mannheim 1991

Goertz, K. C.: »Staub-Plasma-Wechselwirkungen«, in: »Glassmeier/Scholer, 1991, p. 305–330

Gold, T.: »The origin of natural gas and petroleum, and the prognosis for future supplies«, in: »Annual Review of Energy«, Nov. 1985, vol. 10, p. 53 ff.

Gold, T.: »Power from the Earth«, London 1987

Gold, T.: »The deep, hot biosphere«, in: »Proceedings of the National Academy of Sciences«, 1992, vol. 89. p. 6045–6049

Gold, T.: »The Deep Hot Biosphere«, New York 1999

Gold, T., und Held, M.: »Helium-nitrogen-methane-sysematics in natural gases of Texas and Kansas«, in: »Journal of Petroleum Geology«, vol. 10, 1987, p. 415 ff.

Gold, T., und Soter, p.: »The deep-Earth gas hypothesis«, in: »Scientific American«, vol. 242, 1980, p. 154 ff.

Goldberg, R. A., et al.: »Direct observations of magnetospheric electric precipitation stimulated by lightning«, in: »J. Atm. Terr. Phys.«, vol. 48, 1986, p. 293

Gornitz, V., und Fung, I.: »Potential distribution of methane hydrate in the world's oceans«, in: »Global Biogeochem. Cycles«, vol. 8, 1994, p. 335–347

Gruber, J.: »Kalte Fusion und Raumenergie – eine ›neue‹ erneuerbare Energiequelle: Eine einführende Zusammenfassung nach einer Tagung über Kalte Fusion (Vancouver-Bericht)«. Lehrgebiet Statistik und Ökonometrie, FernUniversität Hagen, 10. 6. 1998

Guo, G., und Wang, B.: »Cloud anomaly before Iran earthquake«, in: »International Journal of Remote Sensing«, vol. 29, 7. 4. 2008, p. 1921–1928

Haber, H.: »Unser blauer Planet«, Stuttgart 1965

Haber, H.: »Die Architektur der Erde«, Stuttgart 1970

Halekas, J. p., et al.: »Extreme lunar surface charging during solar energetic particle events«, in: »Geophysical Research Letters«, vol. 34, 30. 1. 2007, doi:10.1029/2006GL028517

Hansen, J., et al.: »Earth's energy imbalance: Confirmation and implications«, in: »Science«, vol. 308, 3. 6. 2005, p. 1431–1435

Harrison, T. M., et al.: »Temperature spectra of zircon crystallization in plutonic rocks«, in: »Geology«, vol. 35, July 2007, p. 635–638

Haywood, J.: »The Natural and Aboriginal History of Tennessee«, Nashville 1823

Herrmann, J.: »Astronomie, die uns angeht«, Gütersloh 1974

Hilgenberg, O. C.: »Vom wachsenden Erdball«, Berlin 1933

Hitchcock, C. H.: »The Albert Coal, or Albertite of New Brunswick«, in: »Amer. J. Sci.«, 2nd Ser. vol. 39, 1865, p. 267–273

Hoefs, J.: »Is biogenic carbon always isotopically ›light‹, is istotopically ›light carbon‹ always of biogenic origin?«, in: »Adv. Organic Geochem.«, 1971, p. 657–663

Hoffe, K. E. A. von: »Geschichte der durch Überlieferung nachgewiesenen natürlichen Veränderungen der Erdoberfläche«, II. Teil, Gotha 1824

Holländer, E.: »Wunder, Wundergeburt und Wundergestalt in Einblattdrucken des 15. bis 18. Jahrhunderts«, Stuttgart 1921

Hsü, K. J.: Klima macht Geschichte«, Zürich 2000

Huc, M.: »Recollections of a Journey through Tartary, Thibet and China, During the years 1844, 1845 and 1846, Band 2«. New York 1852

Hutko, A. R., et al.: »Seismic detection of folded, subducted lithosphere at the core-mantle boundary«, in: »Nature«, 18. 5. 2006, vol. 441, p. 333–336

Ikezi, H.: »Coulomb solid of small particles in a plasma«, in: »Phys. Fluids«, vol. 29, 1986, p. 1764

Ishii, H. A.: »Comparison of Comet 81P/Wild 2 Dust with Interplanetary Dust from Comets«, in: »Science«, vol. 319, 25. 1. 2008, p. 447–450

Jacob, K.-H., et al. (Hrsg.): »Lagerstättenbildung durch Energiepotentiale in der Lithosphäre«, in: »Erzmetall«, vol. 45, 1992, p. 505–513

Javoy, M., et al.: »Carbon geodynamic cycle«, in: »Nature«, vol. 300, 11. 11. 1982, p. 171 ff.

Jensen, P., et al.: »*Bubbling reefs* in the Kattegat: submarine landscapes of carbonate-cemented rocks support a diverse ecosystem at methane seeps«, in: »Marine Ecology Progress Series«, vol. 83, 16. 7. 1992, p. 103–113

Johnson, D. p.: »Phamtom Islands of the Atlantic«, New Brunswick 1944; German: »Fata Morgana der Meere«, München/Zürich 1999

Jordan, P.: »Die Expansion der Erde«, Braunschweig 1966

Judd, A. G.: »Shallow gas and gas seepages: a dynamic process?«, in: Ardus, D. A., und Green, C. D.: »Safety in Offshore Drilling. The Role of Shallow Gas Surveys«, Dordrecht 1990, p. 27–50

Judd, A. G., et al.: »Contributions to atmospheric methane by natural seepages on the UK continental shelf«, in: »Marine Geology«, vol. 140, 1997, p. 427–455

Judd, A. G., und Hovland, M.: »Seabed Fluid Flow«, Cambridge 2007

Kehse, U.: »Mysterium am Meeresgrund«, in: »Bild der Wissenschaft, Ausgabe 6/2000, p. 12–17

Kehse, U.: »Weiß wie Schnee und schwarz wie Ebenholz«, in: »Bild der Wissenschaften«, Online-News 18. 9. 2007

Kelley, Joseph T., et al.: »Giant Sea-Bed Pockmarks: Evidence for Gas Escape from Belfast Bay, Maine«, in: »Geology«, vol. 22, 1994, p. 59

Kelly, P. M., und Wigley, T. M. L.: »Solar cycle length, greenhouse forcing and global climate«, in: »Nature«, vol. 360, 26. 11. 1992, p. 328–330

Kenney, J. F., et al.: »The evolution of multicomponent systems at high pressures: VI. The thermodynamic stability of the hydrogen-carbon system: The genesis of hydrocarbons and the origin of petroleum«, in: »Proceedings of the National Academy of Sciences«, vol. 99, 12. 8. 2002,

p. 10976–10981

Kent, P. E., und Warman, H. R.: »An environmental review of the world's richest oil-bearing region – the Middle East«, in: »Internat. Geol. Congr. 24th Sect.«, vol. 5, 1972, p. 142–152

Kerr, R. A.: »German Super-Deep Hole Hits Bottom«, in: »Science«, vol. 266, 28. 10. 1994, p. 545

Kim, K., et al.: »Methane: a real-time tracer for submarine hydrothermal systems«, in: »EOS«, vol. 64, 1983, p. 724

King, M. B.: »Tapping the Zero-Point Energy«, Kempton 2002; deutsch: »Die Nutzbarmachung der Nullpunktenergie«, Peiting 2003

Kippenhahn, R., und Möllenhoff, C.: »Elementare Plasmaphysik«, Zürich 1975

Kissel, J., und Krueger, F. R.: »Urzeugung aus Kometenstaub«, in: »Spektrum der Wissenschaft«, 5/2000, p. 64–71

Kirby, p. H., und McCormick, J. W.: »Inelastic Properties of Rocks and Minerals:Strength and Rheology«, in: »Handbook of Physical Properties of Rocks, vol. 2«, Boca Raton 1982, p. 151 f., 170

Klauda, J., und Sandler, p. I.: »Global distribution of methane hydrate in ocean sediment«, in: »Energy & Fuels«, vol. 19 (2), 2005, p. 459–470.

Knauth, L. P., und Lowe, D. R.: »Oxygen isotope geochemistry of cherts from the Onverwacht Group (3.4 billion years), Transvaal, South Africa, with implications for secular variations in the isotopic compositions of cherts«, in: »Earth and Planetary Science Letters«, vol. 41, 1978, p. 209–222

Köchling, M.: »Es werde Licht! Die Physik des Universums«, Erkrath 2001

Kozlovsky, Y. A.: »Kola Super-Deep: Interim Results and Prospects«, in: »Episodes. Journal of International Geoscience«, vol. 5, Nr. 4, Dezember 1982

Kraiss, A.: »Geologische Untersuchungen über das Ölgebiet von Wietze in der Lüneburger Heide«, Archiv für Lagerstättenforschung, Heft 23, Berlin 1916

Kravtsov, A. I.: »Inorganic generation of oil and criteria for exploration for oil and gas«, in: »Nefti Gaza«, Kiev 1975, p. 38–48

Kropotkin, P. N.: »Earth's Outgassing and Geotectonics« (Degazatsiia Zemli I Geotektonika), Moskau 1976

Kropotkin, P. N., und Valyaev, B. M.: »Tectonic control of Earth outgassing and the origin of hydrocarbons«, in: »Proc. 27th Internat. Geol. Congr.«, vol. 13, 1984, p. 395–412

Kropotkin, P. N.: »Degassing of the Earth and the origin of hydrocarbons«, in: »Internat. Geol. Rev.«, vol. 27, 1985, p. 1261–1275

Kropotkin P. N.: »Earth's outgassing and genesis of hydrocarbons«, in: »Mendeleev All-Union Chem. Soc. Jour. – Moscow«, (Khimiya) Chemistry,vol. XXXI (5), 1986, p. 540–546 (Russian)

Krug, H.-J., et al.: »Morphological Instabilities in Pattern Formation by Precipitation and Crystallization Processes«, in: »Geol. Rundschau«, vol. 85, 1996, p. 19–28

Krug, H.-J., und Jacob, K.-H.: »Genese und Fragmentierung rhythmischer Bänderungen durch Selbstorganisation«, in: »Z. Dt. Geol. Ges.«, vol. 144, p. 451–460

Krug, H.-J., und Kruhl, J. H.: »Selbstorganisation«, vol. 11 2000, Berlin 2001

Kudryavtsev, N. A.: »Geological proof of the deep origin of Petroleum«, Issledovatel Geologoraz Vedoch, Inst. No. 132, 1959, p. 242–262 (Russian)

Kukuk, P.: »Geologie des Niederrheinisch-Westfälischen Steinkohlengebietes«, Berlin 1938

Kundt, W.: »Astrophysics. A New Approach«, Berlin/Heidelberg/New York 2001, 2. Auf. 2005

L'Haridon, p., et al.: »Hot subterranean biosphere in a continental oil reservoir«, in: »Nature«, vol. 377, 21. 9. 1995, p. 223 f.

Landes, K. K.: »Petroleum Geology of the United States«, New York 1970

Lay, T., und Garnero, E. J.: »Core-mantle boundary structures and processes«, in: Sparks, R. p. J., und Hawkesworth, C. J.: »The State of the Planet: Frontiers and Challenges in Geophysics« (Hrsg.),

American Geophysical Union,

Geophysical Monograph 150, IUGG Volume 19, Washington D.C. 2004, p. 25–41

Lehmann, B.: »Globale chemische Fraktionierungstrends und Lagerstättenbildung«, Berliner Geowissenschaften, Abhandl. A, vol. 167, 1994, p. 57–65

Liesegang, R. E.: »Geologische Diffusion«, Dresden/Leipzig 1913

Lin, R. P., et al.: »Lunar Surface Magnetic Fields and Their Interaction with the Solar Wind: Results from Lunar Prospector«, in: »Science«, vol. 281, 4. 9. 1998, p. 1480–1484

Lin, A., et al.: »Co-Seismic Strike-Slip and Rupture Length Produced by the 2001 Ms 8.1 Central Kunlun earthquake«, in: »Science«, vol. 296, 14. 6. 2002, p. 2015 ff.

Lyell, Ch.: »Principles of Geology«, vol. 1, New York 1872, 11. Ed.

Mac-Donald, I. R.: »Bottom line for hydrocarbons«, in: »Nature«, vol. 385, 30. 1. 1997, p. 389 f.

Mahfoud, R. F., und Beck, J. N.: »Why the Middle East fields may produce oil forever«, in: »Offshore«, April 1995, p. 58–64, 106

Manley, G.: »Central England Temperatures: Monthly Means 1659–1973«, in: »Quat. Journal Roy Meteorol. Soc.«, Ausg. 100, 1974, p. 389–405

Marcus, P. p.: »Prediction of a global climate change on Jupiter«, in: »Nature«, vol. 428, 22. 4. 2004, p. 828–831

Markson, R., und Muir, M.: »Solar Wind Control of the Earth's Electric Field«, in: »Science«, vol. 208, 30. 5. 1980, p. 979–990

Marty, B., und Jambon, A.: »C/He3 in volatile fluxes from the solid Earth: implications for carbon geodynamics«, in: »Earth Planet. Sci. Lett.«, vol. 83, 1987, p. 16–26

Maxlow, J.: »Terra non Firma Earth«, Wroclaw 2005

Meier, G.: »Die deutsche Frühzeit war ganz anders«, Tübingen 1999

Mendelejew, D. I.: »L'origine du petrole«, in: »Revue Scientifique«, 2e Ser., VII, 1877, p. 409–416

Mereschkowsky, K. p.: »Über Natur und Ursprung der Chromatophoren im Pflanzenreiche«, in: »Biol. Centralbl.«, vol. 25, 1905, p. 593–604 und 689 ff.

Merriam, J. C.: »The Fauna of Rancho La Brea«, Memoirs of the University of California, I, Nr. 2, 1911

Meyl, K.: »Elektrische Umweltverträglichkeit, Teil 1«, Villingen-Schwenningen 1966, 3. Ed. 1998

Meyl, K.: »Elektromagnetische Umweltverträglichkeit. Teil 2«, Villingen-Schwenningen 1998, 3. Ed. 1999

Middlehurst, M., und Moore, P.: »Lunar Transient Phenomena: Topographical Distribution«, in: »Science«, vol. 155, 27. 1. 1967, p. 449 ff.

Milkov, V.: »Global estimates of hydrate-bound gas in marine sediments: how much is really out there?«, in: »Earth-Science Reviews«, vol. 66, 2004, p. 183–197

Miller, R. V.: »Bacterial gen swapping in nature«, in: »Scientific American«, Jan. 1998, p. 67 ff.

Moini, R., et al.: »An Antenna Theory Model for the Lightning Return Stroke«, in: »Proc. of the 12th Int. Zurich Symp. on EMC, Zurich, Switzerland, February 18-20«, 1997, p. 149–152, R. Moini, V. A. Rakov, M. A. Uman, and B. Kordi

Moini, R., et al.: »A New Lightning Return Stroke Model Based on Antenna Theory«, in: »J. Geophys. Res.«, vol. 105, 2000, p. 29 693–29 702

Moore, R.: »Die Evolution«, in der Reihe: »Life-Wunder der Natur«, 1970

Mühleisen, R.: »The global circuit and its parameters«, in: Dolezalek, H., und Reiter, R.: »Electrical Processes in Atmospheres«, Darmstadt 1977, p. 467

Müller, F.: »Beobachtungen über Pingos«, Kopenhagen 1959

Mukhtarov, A. p., et al.: »Temperature evolution in the Lokbatan Mud Volcano crater (Azerbaijan) after eruption of 25 October 2001«, in: »Energy Exploration & Exploitation«, vol. 21, 2003, p. 187–207 Impact«, in: »Science«, vol. 310, 14. 10. 2005, p. 270–274

Mumma, M. J., et al.: »Strong Release of Methane on mars in Northern Summer 2003«, in: »Science

Online«, 15. 1. 2009

Murawski, H., und Meyer, W.: »Geologisches Wörterbuch«, Stuttgart 1937, 10. Ed. 1998

Nelson, F. E.: »(Un)Frozen in Time«, in: »Science«, vol. 299, 14. 3. 2003, p. 1673 ff.

Nelson, J. p., und Simmons, E. C.: »Diffusion of methane and ethane through the reservoir cap rock: Implications for the timing and duration of catagenesis«, in: »American Association of Petroleum Geologists Bulletin«, Juli 1995, vol. 79, No. 7, p. 1064–1074

Nemchin, A. A., et al.: »A light carbon reservoir recorded in zircon-hosted diamond from the Jack Hills«, in: »Nature«, vol. 454, 3. 7. 2008, p. 92–95

Ne˘mec, F., et al.: »Spacecraft observations of electromagnetic perturbations connected with seismic activity«, in: »Geophysical Research Letters«, vol. 35, 15. 3. 2008, doi:10.1029/2007GL032517

Neubauer, F. M.: »Die Magnetosphären anderer Planeten im Sonnensystem«, in: »Glassmeier/Scholer, 1991, p. 184–206

Newton, I.: »Optics«, London 1730, 4. Ed.

Nisbet, E. G., und Fowler, C. M. R.: »Some liked it hot«, in: »Nature«, vol. 404, 1. 8. 1996, p. 404 f.

Oesterle, O., und Jacob, K. H.: »Über Lagerstättenbildung durch elektrische Felder«, in: »Zeitschrift der Förderer des Bergbau- und Hüttenwesens an der TU Berlin«, 1994, p. 21–29

Oesterle, O.: »Kann die Geschwindigkeit des radioaktiven Zerfalls künstlich gesteuert werden?«, in: »RFQ- Magnetik«, Sonderausgabe 1996, Rapperswil (Schweiz)

Oesterle, O.: »Evolution der unbelebten und lebenden Substanz der Erde vom Standpunkt der statistischen Chemie«, in: »Idee der Entwicklung in der Geologie«, 1990, p. 121–131

Oesterle, O.: »Goldene Mitte: Unser einziger Ausweg«, Rapperswil am See 1997

Oort, A. H., et al.: »Historical trends in the surface temperature over the oceans based on the COADS«, in: »Climate Dynamics«, 1987 (2), 29–38

Ourisson, G., et al.: »The microbial origin of fossil fuels«, in: »Sci. Am.«, vol. 251, 1984, p. 44–51

Overeerm, I., et al.: »The Late Cenozoic delta system in the Southern North Sea Basin: a climate signal in sediment supply?«, in: »Basin Research«, vol. 13, 2001, p. 293–312

Mumma, M. J., et al.: »Parent Volatiles in Comet 9P/Tempel 1: Before and After Impact«, in: »Science«, vol. 310, 14. 10. 2005, p. 270–274

Schultz, P. H., et al.: »Lunar activity from recent gas release«, in: »Nature«, vol. 444, 9. 11. 2006, p. 184 ff.

Pailu, C. K., et al.: »Methane-rich plumes on the Carolina continental rise: Associations with gas hydrates«, in: »Geology«, vol. 23, 1/1995, p. 89

Paret, O.: »Das neue Bild der Vorgeschichte«, Stuttgart 1964

Parker, E. N.: »Dynamics of the interplanetary gas and magnetic fields«, in: »Astrophys. J.«, vol. 128, 1958, p. 664–675

Parkes, J., und Maxwell, J.: »Some like it hot (and oily)«, in: »Nature«, vol. 365, 21. 10. 1993, p. 694 f.

Pasko, V. P., et al.: »Electrical discharge from a thundercloud top to the lower ionosphere«, in: »Nature«, vol. 416, p. 152 ff.

Paull, C. K., et al.: »Assessing methane release from the colossal Storegga submarine landslide«, in: »Geophysical Research Letters«, vol. 34, 16. 2. 2007, L04601

Pedersen, K. R., und Lam, J.: »Precambrian organic compounds from the Ketilidian of South-West Greenland«, in: »Gronlands Geologiske Unders. Bull.«, 1970, No. 82

Pérez-Peraza, J., et al.: »Solar, geomagnetic and cosmic ray intensity changes, preceding the cyclone appearances around Mexico«, in: »Advances in Space Research«, vol. 42, 3. 11. 2008, p. 1601–1613

Peterson, W.: »Dinosaur tracks in the roofs of coal mines«, in: »Nat. Hist.«, vol. 24 (3), 1924, p. 388

Philippi, G. T.: »On the depth, time and mechanism of origin of the heavy to medium-gravity naphthenic crude oils«, in: »Geochim. Cosmochim«, vol. 41, 1977, p. 33–52

Pickford, M.: »The expanding Earth hypothesis: a challenge to plate tectonics«, in: Scalera/Jacob, 2003, p. 233–242

Potonié, H.: »Entstehung der Steinkohle«, Berlin 1905

Prasser, H.-M.: »Gedanken zur Versorgungssicherheit beim Kernbrennstoff«, Referat von Prof. Horst-Michael Prasser, ETH Zürich, anlässlich der Jubiläumsveranstaltung des Nuklearforums Schweiz am 29. 5. 2008 in Lausanne

Pratt, W. E.: »Towards a philosophy of oil-finding«, in: »Bull. Am. Assoc. Petroleum Geologists 36«, vol. 12, 1952, p. 2231–2236

Price, L. C.: »Organic geochemistry of core samples from an ultradeep hot well«, in: »Chem. Geol.«, vol. 37, p. 215–228

Raeder, J.: »Grundlagen der numerischen Plasmasimulation«, in: »Glassmeier/Scholer, 1991, p. 331–352

Raffles, T. p.: »The History of Java«, 1817, vol. 1

Rahmstorf, p., und Schnellnhuber, H. J.: »Der Klimawandel«, München 2006, 4. Ed. 2007

Rakov, V. A., und Uman, M. A.: »Lightning. Physics and effects«, Cambridge 2003

Ranada, A. F., et al.: »Ball lightning as a force-free magnetic knot«, in: »Phys. Rev. E 62«, 2000, p. 7181–7190

Ranzani, G.: »Astronomie«, Klagenfurt 2001

Raulin, F.: »Planetary science: Organic lakes on Titan«, in: »Nature«, vol. 454, 30. 7. 2008, p. 587 ff.

Ritger, p., et al.: »Methane-derived authigenic carbonates formed by subduction-induced porewater expulsion along the Oregon-Washington margin«, in: »Geol. Soc. Am. Bull.«, vol. 98, p. 147–156

Robinson, R.: »Duplex Origin of Petroleum«, in: »Nature«, vol. 199, 13. 7. 1963, p. 113 f.

Robinson, R.: »The Origins of Petroleum«, in: »Nature«, vol. 212, 17. 12. 1966, p. 1291–1295

Rompe, R.: »Der vierte Aggregatzustand«, Leipzig/Jena 1957

Rudenko, A. P.: »Evolutions-Chemie und natürlich-historischer Standpunkt im Problem des Entstehen des Lebens«, in: »Journal chem. Gesellschaft«, UdSSR, vol. 25/4, p. 390–484, Moskau 1980

Sandberg, C. G. p.: »Ist die Annahme von Eiszeiten berechtigt«, Leiden 1937

Sawenkow, W. J.: »Neue Vorstellungen über das Entstehen des Lebens auf der Erde«, in: »Wyschtscha schkola«, 1–231, Kiew 1991, in Russisch (zitiert in Oesterle, 1997)

Scalera, G., und Jacob, K.-H.: »Why expanding Earth?«, Rom 2003

Schidlowski, M., et al.: »Precambrian sedimentary carbonates: Carbon and oxygen isotope geochemistry and implications for the terrestrial Oxygen budget«, in: »Precambrian Research«, vol. 2, 1975, p. 1–69

Schidlowski, M.: »Die Geschichte der Erdatmosphäre«, in: »Die Dynamik der Erde«, Heft: »Spektrum der Wissenschaft: Verständliche Forschung«, Heidelberg 1987, 2. Ed. 1988

Schimper, A. F. W.: »Über die Entwicklung der Chlorophyllkörner und Farbkörper«, in: »Botanische Zeitung«, vol. 41, 1883, Sp. 105–120, 126–131 und 137–160

Schlegel, K.: »Das Polarlicht«, in: Glassmeier/Scholer, Mannheim 1991, p. 164–183

Schönwiese, C.-D.: »Klimaänderungen. Daten, Analysen, Prognosen«, Berlin 1995

Scholer, M.: »Die Magnetopause der Erdatmosphäre«, in: Glassmeier/Scholer 1991, p. 119–138

Schrödinger, E.: »Was ist Leben? Die lebende Zelle mit den Augen des Physikers betrachtet«, München 1951, 2. Auflage

Schultz, P. H., et al.: »Lunar activity from recent gas release«, in: »Nature«, vol. 444, 9. 11. 2006, p. 184 ff.

Schwarzbach, M.: »Das Klima der Vorzeit«, Stuttgart 1993

Schweigert, I. V., und Schweigert, V. A.: »Forces acting on a coulomb crystal of microparticles in plasma«, in: »Journal of Applied Mechanics and Technical Physics«, vol. 39, Nov. 1998, p. 825–831

Schwenn, R.: »Der Sonnenwind«, in: Glassmeier/Scholer, 1991, p. 17–46

Scott, H. P., et al.: »Generation of methane in the Earth's mantle: In situ high pressue-temperature measurements of carbonate reduction«, in: »Proceedings of the National Academy of Sciences«, vol. 101, 28. 9. 2004, p. 14 023–14 026

Sereno, P. C.: »The Evolution of Dinosaurs«, in: »Science«, vol. 284, 25. 6. 1999, p. 2137–2147

Shcheka, p.: »Carbon in the Earth's Mantle«, Dissertation an der Eberhard-Karls-Universität Tübingen, 2006

Shiga, D.: »Fizzy water powered ›super‹ geysers on ancient Mars«, in: »New Scientist« Online-Dienst, 17. 3. 2008

Silberg, P. A.: »Ball Lightning and Plasmoids«, in: »J. Geophys. Res.«, vol. 67, Nr. 12, 1962

Singer, p.: »The Nature of Ball Lightning«, New York 1971

Sokolov, V. A., et al.: »The origin of gases of mud volcanoes and the regularities of their powerful eruptions«, in: »Adv. Organic Geochem.«, 1968, p. 473–484

Sokoloff, W.: »Kosmischer Ursprung der Bitumina«, in: »Bull. Soc. Imp. Natural Moscau«, Nouv. Ser. 3, 1889, p. 720–739

Stallard, T., et al.: »Jovian-like aurorae on Saturn«, in: »Nature«, vol. 453, 19. 6. 2008, p. 1083 ff.

Stehn, C. E.: »The geology and volcanism of the Krakatau group«, in: »Proc. Fourth Pacific Science Congress, Batavia 1929«, p. 1–55

Storey, M., et al.: »Paleocene-Eocene Thermal Maximum and the Opening of the Northeast Atlantic«, in: »Science«, vol. 316, 24. 4. 2007, p. 587 ff.

Strugov, A. p.: Die Explosion eines Hydrolakkolithen, in: »Natur 6«, Moskva 1955

Stutzer, O.: »Die wichtigsten Lagerstätten der Nicht-Erze, I. Erdöl«, Berlin 1931

Su, H. T., et al.: »Gigantic jets between a thundercloud and the ionosphere«, in: »Nature«, vol. 423, 26. 6. 2003, p. 974 ff.

Subarya, C., et al.: »Plate-boundary deformation associated with the great Sumatra-Andaman earthquake«, in: »Nature«, 2. 3. 2006, vol. 440, p. 46–51

Suess, E., und Bohrmann, G., 2002. Brennendes Eis: Vorkommen, Dynamik und Umwelteinflüsse

Suess, E.: »Das Antlitz der Erde« (4 Bände), Leipzig 1885/1909

Suess, E., und Bohrmann, G.: »Brennendes Eis – Vorkommen, Dynamik und Umwelteinflüsse von Gashydraten«, in: Wefer, G. (Hrsg.): »Expedition Erde«. Beiträge zum Jahr der Geowissenschaften 2002, p. 108–116

Suess, H. E.: »Radiocarbon Concentration in Modern Wood«, in: »Science«, vol. 122, 2. 9. 1955, p. 415 ff.

Supan, A.: »Grundzüge der physischen Erdkunde«, Leipzig 1916

Svensmark, H.: »Cosmic rays and Earth's Climate«, in: »Space Science Reviews«, vol. 93, 2000, p. 155–166

Swain, M. R., et al.: »The presence of methane in the atmosphere of an extrasolar planet«, in: »Nature«, vol. 452, 20. 3. 2008, p. 329 ff.

Szewzyk, U., et al.: »Thermophilic, anaerobic bacteria isolated from a deep borehole in granite in Sweden«, in: »Proceedings of the National Academy of Sciences«, vol. 91, 1. 3. 1994, p. 1810–1813

Tesla-Museum: »Nikola Tesla (1956)«, Whitefish 1956

Thorson, R. M., et al.: »Geologic evidence for a large prehistoric earthquake in eastern Connecticut«, in: »Geology«, 14, Boulder 1986

Tollmann, A. und E.: »Und die Sintflut gab es doch«, München 1993

Torkar, K., et al.: »An experiment to study and control the Langmuir sheath around INTERBALL-2«, in: »Annales Geophysicae«, vol. 16, 9/1998, p. 1086–1096

Transehe, N. A.: »The Siberian Sea Road«, in: »The Geographical Review«, vol. 15, 1925, p. 375

Ukraintseva, V. V.: »Vegetation Cover and Environment in the Mammoth Epoch in Siberia«, Hot Springs 1993

Vereshchagin, N. K., und Baryshnikov, G. F.: »Paleoecology of the Mammoth Fauna in the Eurasian Arctic«, New York 1982

Vogel, A.: »Die Kern-Mantel-Grenze: Schaltstelle der Geodynamik«, in: »Spektrum der Wissenschaft«,

11/1994, p. 64–72

Vogel, K.: »The Expansion of the Earth – An Alternative Model to the Plate Tectonics Theory«, in: »Critical Aspects of the Plate Tectonics Theory«; vol. II, »Alternative Theories«, Athens 1990, p. 14–34

Vogt, P. R., et al.: »Methane-Generated (?) Pockmarks on Young, Thickly Sedimented Oceanic Crust in the Arctic: Vestnesa Ridge, Fram Strait«, in: »Geology«, vol. 22, 1994, p. 255

Volland, H.: »Der Plasmazustand der Atmosphäre«, in: Glassmeier/Scholer, 1991, p. 284–304

Vollmer, A.: »Sintflut und Eiszeit«, Obernburg 1989

Walther, J. W.: »Geschichte der Erde und des Lebens«, Leipzig 1908

Wanlass, H. R.: »Studies of field relations of coal beds«, in: »Second Conference on the Origin and Constitution of Coal«, Nova Scotia 1952, p. 148–180

Watson, D. M., et al.: »The development of a protoplanetary disk from its natal envelope«, in: »Nature«, vol. 448, 30. 8. 2007, S, 1026 ff.

Wehlan, J. K.: The dynamic migration hypothesis, in: Sea Technology, September 1997, p. 10 ff.

Weiß, E.: »Littrow, Wunder des Himmels«, Berlin 1886, 7. Ed.

Welhan, J. A., und Craig, H.: »Methane, hydrogen and helium in hydrothermal fluids at 21 N on the East Pacific Rise«, in: Rona, P. A.: »Hydrothermal process at Seafloor Spreading Centers«, 1983, p. 391–409

Wells, D. R.: »Dynamic Stability of Closed Plasma Configurations«, in: J. Plasma Physics, 1970, vol. 4, p. 654

Whelan, J. K., et al.: »Organic geochemical indicators of dynamic flow process in petroleum basins«, in: »Advances in Organic Chemistry«, vol. 22, 1993, p. 587 ff.

Whelan, J. K.: »The dynamic migration hypothesis«, in: »Sea Technology«, September 1997, p. 10 ff.

Wibberenz, W.: »Die Kosmische Strahlung im Sonnensystem«, in: »Glassmeier/Scholer, 1991, p. 47–76

Wiegand, G.: »Fossile Pingos in Mitteleuropa«, Würzburg 1965

Wilkening, L.: »Carbonaceous chondritic material in the solar system«, in: »Naturwissenschaften«, vol. 65, 1978, p. 73–79

Willson, R. C., und Hudson, H. p.: »The Sun's Luminosity Over a Complete Solar Cycle«, in: »Nature«, vol. 351, 2. 5. 1991, p. 42 ff.

Zhang, T., et al.: »Statistics and characteristics of permafrost and ground ice distribution in the Northern Hemisphere«, in: »Polar Geography«, vol. 23(2), 1999, p. 147–169

Zillmer, H.-J.: "Darwin's Mistake", Enkhuizen 2002, 2nd Ed. 2003, original German title: "Darwins Irrtum", Munich 1998, 10th Ed. 2011

Zillmer, H.-J.: "Mistake Earth Science", Victoria 2007, original German title: "Irrtümer der Erdgeschichte", Munich 2001, 5th Ed. 2008

Zillmer, H.-J.: "Dinosaurier Handbuch" (Dinosaur Handbook: not yet published in English), Munich 2002

Zillmer, H.-J.: "The Columbus-Mistake", not yet translated in English, German title: "Kolumbus kam als Letzter", Munich 2004, 3rd Ed. 2009

Zillmer, H.-J.: »The Human History Mistake«, Victoria 2009, original German title: »Evolutions-Lüge«, München 2005, 5. Ed. 2010

Zillmer, H.-J.: »Die Erde im Umbruch«, München 2011, not yet published in English

Zillmer, H.-J.: »Kontra Evolution«, German DVD-Video, Solingen 2007, 3. Ed. Rottenburg 2009

Zimmermann, M. R., und Tedford, R. H.: »Histologic Structures Preserved for 21.300 Years«, in: »Science«, vol. 194, 8. 10. 1976, p. 183 f.

Index

1, 2 The author at a borehole on the gas and oil fields north of Dallas with a sample of the lifted oil.

3, 4 A nearby borehole is plugged with microbes.

5 The W. H. Badget No. 1 blew in on September 24, 1922 spewing salt water and nonflammable gas almost 40 feet above the derrick. The salt water gushed out at a rate of 10.000 barrels a day, coating everything with salt and meaking it look like a winter snow scene.

6, 7 The active mud volcanoes putt out gas and oily mud in Java, Indonesia, consistently since May 2006

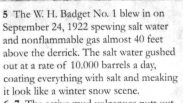

8 Earth fissures and small mud volcano appeared during the Nihonkai-Chubu earthquake in Japan 1983. **Insert**: Schematic sketch of hydrocarbons which migrate and diffuse upward, however, but easier to rise through fault zones (white arrows). Oil collects in "traps".

9 At these earthquakes these "pingos" also arose, scatterd in a paddy-field.

10 From a 100 meters below the surface mud volcano located in the Caspian Sea shot a several mile-high gas flame, which stabilized at a height of 500 meters. It should have been ejected some 300 million cubic meters of gas. The photo was taken from Baku on November 15th, 1958 (Geological institute of Azerbaijan). **Insert**: lateral extension of the flame (120 meters).

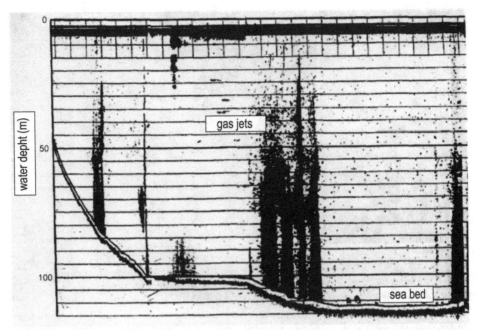

11 During echo sounder work southeast of the Greek island of Milo two earthquakes happened in March 1992. The surface of the sea started to "boil" and it were registered 99 gas jets. Three months later 60 were always still active (Dando et al., 1995). From 34 square kilometres seabed around Greek islands (Kos, Lesvos, Santorini, Methana, etc.) arises gas.

12 An intense eruption of gasses which mainly consisted of methane next to a drilling platform (by oval marked) in the Norwegian Sea from a depth of 240 meters. (Judd/Hovland, 2007, S. 365).

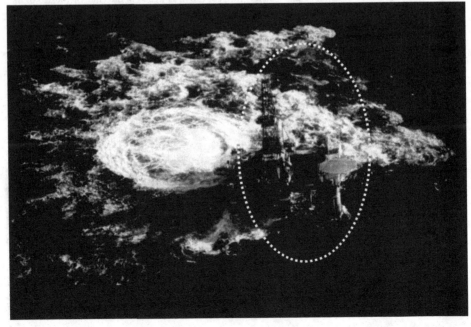

261

13 13 White and Black smokers (sea vents) are rmineral tubular structures on the seabed from which flows of hot water with minerals, dissolved salts and sulfides and carbon dioxide or methane. **Insert**: This white smoker at Eifuku volcano near the Mariana Islands in the Pacific come from fat bubbles. After more than 20 years, such tubes are often "concreted" over closing the opening.

14 Artistic portrayal of "bubbling reefs" (after: Chr. W. Hanson). Columnar sub-marine structures – from which bubbles escape – occur in places where methane seeps upward through the marine sediments. Partly occur due to oxidation of methane to form carbon dioxide, which chemically are formed carbonates. This "concrete" solidifies the sediments of the seafloor and as a result to reduce the pressure there arise hollow stone columns in which hot water and minerals flow upward. Such structures has been detected at the Oregon subduction zone at depth of 2,000 meters (Ritgen, 1987), as well as in the Gulf of Mexico or in shallow marine sediments, as in the North Sea (Jensen, 1992).

15-17 The up to ten feet high are the hollow carbonate-structures of mineral origin at "Pobiti Kamani", located in Varna Province, Bulgaria. It is suggested that these rock formations are organic origin or petrified trees as they look like. But it is testimony to an ancient seabed, where was seeping out methane.

18 In the Kattegat off the Danish coast occur carbonate-tubes without "foundation" sticking in the sandy sea floor. Those reach a height of up to four meters, with up to twelve meters water depth. These structures are often mistakenly viewed as coral reefs. However, these are inorganic carbonate formations as a result of methane escaping from the seabed. By sonar images of the Danish island north Hirsholmene gases were detected, marking out of the submarine columnar structures (escape through arrows) after Jensen, 1992. **Insert**: Up to 20 m high tubular structures Ikaite, a calcium carbonate phase, in the Ikka Fjord, Greenland (Buchardt et. Al., 2001).

19 *An exposed pingo: Such an ice body can develop by the Joule-Thomson effect, if rising gas or liquid is forced through a porous plug (soil). The adiabatic (no heat exchanged) expansion of a gas may be carried out in a number of ways. Methane oxidized and the resulting water freezes. Then it will be formed an earth covered block of ice. If the pressure is too high ice volcanoes are created, as they are found not only in permafrost regions. Also this pingo has a crater above, now filled with water, escaped through the gas, sometimes suddenly.*

20 *A 50 m high with pingo ice core in the Mackenzie Delta in the Northwest Terretorien, Canada*

31 *A large Pingo 35 km east of Longyearbyen on Svalbard, Spitsbergen*

22 In Montana, near the border to Canada, there is a swarm of 300 about 0.38 to 2.4 km elongated drumlins: The 240 square kilometres large "Eureka Drumlin Field". Also interesting is the lateral horizontal displacement

23 A drumlin west of Calgary, Canada, where are these landscape forms common phenomena.

24 An exposed drumlin withthout ice core in Scotland showing a spherical, towards the center upright stratification.

25 The picture shows pingos as seldom recognized form of landscape on the west coast of Svalbard, Spitzbergen: Wall rings from boulders are surrounding fine grained

26 Volcanic lightning on Mount Rinjani in Indonesia. Photo: Oliver Spalt, 1995.
27 Eruption of the mud volcano Lokbatan in Azerbaijan in 2001 with a 400 meter-high flame: The black smoke indicates that combustion of higher hydrocarbons.
28 Burning hydrocarbons in Azerbaijan (top) and the mud volcanic Yangnü Xin Hu Yanchao, China (Photo: Wolfgang Odendahl).

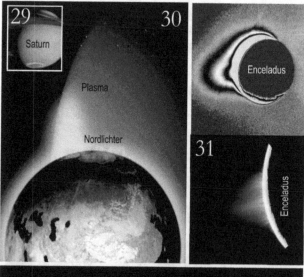

29 The Cassini spacecraft photographed 2005, the first ring-shaped auroras at Saturn's south pole.

30 A plasma fountain of oxygen, helium and hydrogen over the North Pole. Northern Lights are phenomena. Image: NASA.

31 The visible fountains at the south pole of Saturn's moon Enceladus. Upper picture: Giant energy processes clarified by image editing. NASA / JPL, 2005.

32 Jupiter's moon Io is the actives' volcanic planetary body. Icy volcanic Eruptions consist of sulphur and sulphur dioxide. Insert: With liquid sulphur-filled volcanic vent Tupan Patera. Diameter 75 km.

A German Bestseller:

In his first bestseller, *Darwin's Mistake*, which has meanwhile been translated into twelve languages, the independent private scholar and consulting civil engineer Hans-Joachim Zillmer has proved that there were megafloods processes and cataclysms but no macroevolution, whereas *Mistake Earth Science* presents a global revolution in the truest sense of the word in an exciting format. It's a lot of fireworks of facts and evidence, but also "live" excavations with the author and visits to all continents permit a radical revision of previously imagined worlds in favour of newer, trend-setting models of thought.

The various alternative models of thought that were previously introduced by the author have meanwhile been scientifically confirmed: a shift in the Earth's axis of at least 20 degrees at the time of the dinosaurs, and also some cataclysms - the "global" deluge? - at the time of human life, because genetic investigations have shown that mankind had once almost died out. Or also, for example, that bird simply doesn't descend from dinosaurs. Here, the author also documents each piece of evidence that is to be checked out against the theory of evolution, which thereby ends as a beautifully fictional fairy-tale.

In this book, geological and geophysical scenarios and facts are introduced that are still widely relatively unknown. How does one explain that what was previously regarded as the "cradle" of evolutionary theory is geologically too young? Because the Galapagos Islands are only a few million years old and do not stem from the time of the dinosaurs as Charles Darwin erroneously suspected. Why were the largest hippos ever swimming in the rivers of Central European, and this was during the great ice age as well? Dinosaurs once lived on all continents, even on Spitsbergen and in Alaska and at the South Pole, since there was a tropical climate from pole to pole. And now even today, the polar cap ice is melting down dramatically rapidly. In a few thousand years, there will not be any more ice, just as it was during the Mesozoic era, at both poles. More and more new finds of the same dinosaur species on almost all continents, but also on both sides of the Atlantic, have put continental drift in the form in which it has been promoted until now into question. On the basis of the most recent NASA research, there could have been geo-electrical events on Earth that were previously not considered possible: is continental drift an incorrectly interpreted event? Oceans virtually devoid of water, a Mediterranean that dried up, even the presence of hippos on islands such as Cyprus and other phenomena are scenarios that are not to be explained by our world view and currently observable events. The incredible claim advanced by the best-selling author of this book, that the Amazon formerly originated out of the Sahara and flew into the Pacific, was already confirmed scientifically while the book was in press. Whoever is interested in the origins and development of our planet, as well as our present biosphere, won't be able to extract himself from the fascination of this logical demonstration of evidence and will see the Earth's history with completely new eyes.

272 pages; quality trade paperback (softcover); catalogue #07-0105, Trafford Publishing
ISBN 1-4251-1735-X; **US$ 34.90**, C$ 40.14, EUR 27.21, £ 18.04

THE SENSATIONAL NEW DISCOVERY
OF THE HUMAN HISTORY

Our ancestors didn't live in trees, and apes never turned into humans. In The Human History Mistake, German bestselling author Hans-Joachim Zillmer has compiled factual material and empirical facts from all over the world proving that Charles Darwin's evolution theory is a myth.

For more than thirty years, Zillmer has concentrated on investigating contentious findings and inconsistencies in the images of the world, recording numerous sensational discoveries and showing that documenting the anthropogenesis must be changed. In The Human History Mistake, Zillmer points to numerous finds from the Stone Age that are far younger than previously thought. The skulls of Neanderthal man and of people from the Paleolithic age must be made "younger" by as much as 27,000 years to the age of a few thousand or even hundreds of years.

This science book rejects the ideas of macroevolution, but instead demonstrates that microevolution plays a much larger role in the creation of new species. Accompanied by sixty-nine photos and forty-nine illustrations, The Human History Mistake shows that the history of mankind must be rewritten.

Product Details

ISBN-13. 9781426923524 • ISBN 10: 142692352X • Publisher: Trafford Publishing
Publication: 1/21/2010 • Pages: 252 • Size:: 7.5 x 9.25 • Perfect Bound Softcover (B/W)
US$ 21,90